AI for Business

A Practical Guide to Harnessing the Power of AI for Your Business

Matt Greenwood

Contents

Introduction

Welcome to cutting-edge business transformation. If you've picked up this book, you're probably curious about how artificial intelligence is reshaping business. Let's start with a bold statement: AI is not just another tech trend. It's a fundamental shift in how we approach business problems, make decisions, and create value. It's like electricity was to the Industrial Revolution—rewriting the rules of competition across industries.

So, what exactly is the role of AI in modern business? In short, it augments human capabilities without necessarily replacing them. AI excels at processing vast amounts of data, recognizing patterns, and making predictions. It can automate routine tasks, freeing up humans to focus on more creative and strategic work. It can provide insights that would be impossible for humans to derive on their own, helping us make better decisions. It can also enable new products, services, and business models that were previously unimaginable.

But let's take a moment to debunk some myths. Despite what sci-fi movies might have you believe, AI isn't a magical solution to all problems, nor is it a sinister force plotting world domination. AI can't think like a human. It doesn't have consciousness or emotions. What we call "artificial intelligence" today is narrow or weak AI-systems that merely perform specific tasks. They can be incredibly good at these tasks, even surpassing human performance, but they can't transfer that ability to other areas.

AI also can't be creative in the same way humans are. While AI can generate art or music from patterns it learned, it doesn't have original ideas or understand the meaning behind its creations. It can't empathize, build relationships, or lead teams—all aspects of business that will remain firmly in the human domain.

Moreover, AI isn't infallible. It can make mistakes, and those mistakes can be amplified without adequate monitoring. AI systems can also perpetuate favoritism present in their training data, leading to unfair or discriminatory outcomes without proper implementation.

So, what can AI do? It can process and analyze vast amounts of data at incredible speeds. It can recognize patterns and make predictions based on historical data. It can automate repetitive tasks with high accuracy. It can also adapt and improve its performance over time through machine learning.

Throughout this book, we'll explore the different types of AI and their business applications through examples and case studies of successful AI implementations across various industries. We'll discuss the ethical considerations and potential pitfalls of AI adoption. Most importantly, we'll provide practical guidance on how to develop and implement an AI strategy that aligns with your business goals. You'll learn how to identify opportunities for AI in your organization, how to build or buy AI solutions, and how to manage the organizational changes that come with AI adoption. We'll cover data preparation, model selection, deployment, monitoring, and more.

But this isn't just a technical manual—the human side of AI adoption—how to build AI literacy in your organization, how to manage the impact on your workforce, and how to foster AI-ready culture, will all come up. By the end of this book, you'll have a comprehensive understanding of AI's role in business and a roadmap for leveraging AI in your own organization through valuable insights and actionable advice.

Part I

Foundations of AI

Chapter One

Understanding Artificial Intelligence

The budding, eerie, fascinating world of artificial intelligence (AI) is where science fiction meets reality. AI is the magic wand of the modern era, transforming everything it touches. Chatbots understand your mood, and self-driving cars take you on a ride around the city. AI is all around, making things easier, smarter, and a whole lot more interesting while raising some complex and potentially morally ambiguous questions. Love it or hate it, or just want to see where it all goes, it's hard to remain neutral about it.

This chapter will unravel the mysteries of AI through its core concepts and terminologies. We'll explore the different types of AI, from the specialized Narrow AI that powers your favorite apps to the ambitious dreams of General AI. We'll also take a trip down memory lane through the history of AI, from its humble beginnings to the groundbreaking advancements that are shaping the future.

Key Terminologies in AI

Learning about AI means learning the terminology first. Let's start with machine learning (ML). Think of ML as the brain behind AI. It teaches a computer to learn from data instead of programming it for specific tasks. It's like trying to teach a dog to fetch a ball. You could manually show the dog each step, or you could just throw the ball and let the dog figure it out through trial and error. ML is the computer figuring it out from examples: many examples of how something looks, reads, and works until it can recognize patterns on its own.

Next up are Neural Networks. These are the superstars of AI, inspired by the human brain. To visualize an AI neural network, consider a web of interconnected neurons that can process information in layers, much like what your own body has. They're great at recognizing patterns, like identifying your face in a photo or understanding your voice. Think of them as layers of filters that process data, each layer capturing different features. For example, the first layer might recognize edges in an image, the next layer might recognize shapes, and so on, until it knows that it's looking at a picture of you or your cat.

Natural language processing (NLP) is another major player in AI. It gets machines to understand, interpret, and respond to human language—naturally spoken, not in a strictly structured form of computer code, as traditional programs would. It's like teaching your computer to speak and understand humans. NLP powers chatbots, translation services, virtual assistants, and those handy autocomplete suggestions in your texts. It's a mix of linguistics and AI, and it's why we can have semi-decent conversations with our devices.

Deep learning is a fancy subset of machine learning. It's layers upon layers of neural networks working together. Deep learning is why we have amazing things like self-driving cars and super-accurate voice assistants. Deep learning handles massive amounts of data and complex computations. The technology is everywhere, from advanced robotics to predicting diseases from medical images.

Computer vision lets machines see and understand the world through images and videos. From tagging your friends in photos to guiding autonomous drones, computer vision gives machines the ability to see. Computer vision is used in security systems, healthcare (like analyzing X-rays), multimodal chatbots (those that can do more than process text), and your car's lane-keeping system, among other things.

Generative AI is the creative side of AI and is closer to what most people imagine when they think of artificial intelligence. It can create content like text, images, music, and video. It's why we have things like AI-generated art and even some of those bizarre deepfake videos. Generative AI can take humongous volumes of data and create something new. For instance, it can write a new song in the style of your favorite artist or generate a realistic-looking human face from scratch. It's both fascinating and a bit spooky, especially if used irresponsibly, but the potential is undeniable.

When we think of AI, we often picture robots. But that's not entirely accurate. AI can exist without robots, and very basic "robots" can exist without AI. However, robotics is where AI meets physical machines. These robots can perform tasks ranging from simple repetitive actions in manufacturing to complex surgeries. Think about those robotic vacuum cleaners or drones delivering packages—that's AI in the physical world, making our lives easier and more efficient.

Reinforcement Learning is a type of ML where an agent learns by interacting with its environment and receiving rewards or penalties. It's like training a puppy with treats and scolding. Over time, the AI learns the best strategies to achieve its goals. This technique is behind many game-playing AIs, such as those that can beat human champions in chess, and it's being used to develop more sophisticated autonomous systems.

Types of AI

'Narrow AI, also known as Weak AI, is the kind of AI that's everywhere today. Narrow AI is designed to perform one specific task really well, but that's about it. You can consider it talented at trivia or solving a Rubik's Cube in seconds, but it might struggle with anything outside its specialty.

For example, take Siri or Alexa. They excel at understanding your voice commands, playing your favorite tunes, setting reminders, and sometimes searching the web. But ask them to do something they're not programmed for, like giving you life advice or writing a novel, and they fall short. Similarly, think about the recommendation systems on Netflix or Spotify. These systems analyze what you've watched or listened to and suggest new movies, shows, or songs you might enjoy. They're brilliant at guessing your next binge-watch or your new favorite song but aren't going to help you solve a math problem or plan your next vacation.

Narrow AI also includes the spam filters in your email that keep pesky unwanted messages out of your inbox. They're fantastic at identifying spam using patterns and keywords but aren't going to read and respond to your emails for you. Self-driving cars are another example. They navigate roads, recognize obstacles, and drive safely, but they're not going to cook your dinner or manage your calendar.

Narrow AI, like ChatGPT and Copilot (also based on ChatGPT), is everywhere, making lives easier by handling specific tasks more efficiently than we could. Each system is limited to its specialized function. These AI systems are great at generating text, helping with coding, and answering questions, but they don't go beyond their specific functions. Just like email spam filters and self-driving cars, they excel in their own domains but won't be cooking your dinner or managing your calendar.

Now, let's talk about the flip side: General AI, or Strong AI. This is the kind of AI seen in 'science fiction. General AI refers to a machine with broad cognitive abilities, much like a human. It can learn, "think," and apply knowledge to a wide range of tasks. It's an AI that is as good at cooking a gourmet meal as it is at diagnosing diseases, writing poetry, or figuring out your next business strategy. That might be a bit of an exaggeration, but it is the ultimate goal of Artificial General Intelligence (AGI)

In theory, General AI would be able to switch from solving complex math problems to giving you life advice, all while baking a perfect soufflé. It's the holy grail of AI research, but we're not there yet. Scientists are working on it, but for now, General AI remains in the realm of movies and books.

Achieving General AI is incredibly complex. It involves creating machines that not only mimic human intelligence but, in a way, truly *have* intelligence. They must possess the ability to generalize learning and apply it across various domains. It requires an understanding of context and nuance and the ability to transfer knowledge from one situation to another seamlessly.

Brief History and Evolution of AI

The story of AI starts off in the mid-20th century. It's the 1950s, a time when computers were the size of entire rooms and had less power than some modern calculators. Yet, some brilliant minds were already dreaming about creating machines that could think and learn like humans. The term "Artificial Intelligence" was coined in 1956 during a workshop at Dartmouth College. computer scientists, including the legendary John McCarthy, got together to brainstorm the future of intelligent machines. Their goal was ambitious: to explore the possibility of machines that could perform tasks that normally require human intelligence and go beyond number-crunching.

Fast forward to the 1960s, when the first real attempts at creating AI took place. Researchers developed early proto-AI programs that could solve algebra problems, prove theorems, and sometimes understand a bit of natural language. One of the first programs, ELIZA, created by Joseph Weizenbaum in 1966, mimicked a psychotherapist by responding to typed statements.

'ELIZA used a technique called pattern matching to identify keywords in the user's input and then responded with pre-programmed phrases. For example, if you said, "I'm feeling sad," ELIZA might reply, "Why do you think you are feeling sad?" It wasn't really understanding your emotions, but it gave you the illusion of understanding, which was revolutionary at the time.

The impact of ELIZA was huge. It demonstrated that computers could engage in simple dialogues, sparking interest and excitement in the field of NLP.' Other early NLP programs followed, like PARRY, which simulated a paranoid schizophrenic patient, and SHRDLU, which could understand

and respond to commands in a virtual world made up of blocks. These early efforts laid the groundwork for today's sophisticated chatbots and virtual assistants.

The 1970s, however, were a bit of a rollercoaster for AI. The initial hype started to fizzle out as researchers hit roadblocks. It turned out that creating intelligent machines was way harder than they had anticipated. Funding dried up, and this period became known as the first "AI winter", a time when interest and investment in AI research cooled down significantly.

There were some major technical hurdles. Early AI systems were incredibly limited. They could do specific tasks but struggled with anything requiring broader understanding or adaptability. These systems relied heavily on rule-based programming, which meant they needed extensive hand-coding for every possible scenario they might encounter. It was like trying to teach a robot to play chess by programming every potential move it could ever make—tedious and not very effective.

On top of the technical challenges, there was a serious funding drought. Governments and private investors, who at first happily poured money into AI research, started to lose faith. As often happens, they were expecting quick, revolutionary breakthroughs, but AI progress was much more incremental. When these breakthroughs didn't materialize as fast as they hoped, the money dried up. Researchers found themselves in a tough spot, with ambitious ideas but no resources to bring them to life.

The research community had to adapt to survive. Many AI researchers shifted their focus to more practical and achievable goals, such as developing software for existing technologies. This period also saw a greater emphasis on academic research rather than commercial applications. It was a time of regrouping and rethinking strategies, laying the groundwork for the future AI revival.

The 1980s brought a resurgence of interest. This era saw the rise of expert systems, which could mimic the decision-making of human experts. These systems were used in fields like medicine and finance, helping doctors diagnose diseases and aiding financial analysts in making investment decisions. The development of these systems renewed enthusiasm and funding for AI research.

Expert systems work by combining a knowledge base of human expertise with an inference engine that applies this knowledge to solve problems. One of the early successes was MYCIN, an expert system that could diagnose bacterial infections and recommend antibiotics. MYCIN was able to outperform even some human doctors in its task, showing the potential of expert systems.

Another notable example was XCON, also known as R1, developed by Digital Equipment Corporation (DEC). XCON could configure orders for new computer systems, which involved a complex set of rules and dependencies. Before XCON, configuring these orders was a time-consuming and error-prone manual process. XCON automated it, significantly improving accuracy and saving DEC millions of dollars.

These successes reignited interest in AI, both from a research and a funding perspective. The development of expert systems showed that AI could provide real, tangible benefits to industries. This revival was fueled by a more pragmatic approach to AI. Rather than trying to create an all-inclusive thinking machine (what we eventually came to call AGI), they instead focused on solving specific problems. That was the birth of what is now called Narrow AI.

The resurgence of AI in the 1980s was also bolstered by advances in computing power. Personal computers have become more powerful and affordable, making sophisticated AI applications more feasible. This period laid the foundation for the AI boom of today, proving that sometimes, focusing on practical, incremental improvements can pave the way for groundbreaking advancements.

Then came the 1990s and the dawn of the internet, which opened up new possibilities for AI. Computers were becoming more powerful, and data was more abundant. AI research flourished again, leading to significant breakthroughs. In 1997, IBM's Deep Blue made history by defeating world chess champion Garry Kasparov in a six-game match.

Deep Blue was no ordinary chess computer. It could evaluate 200 million positions per second. An immeasurably impressive feat. It had already faced Kasparov in 1996 and lost, but the rematch in 1997 was different. The final match was a six-game series. Kasparov won the first game, but Deep Blue came back strong, winning the second game. The remaining games ended in draws, except for the final game, in which Deep Blue won, securing its place in history.

The strategies were intense. Deep Blue used brute-force computing power to evaluate all possible moves, while Kasparov relied on human intuition and experience. The victory was a wake-up call to the world, showing that machines could compete with and outperform humans in complex intellectual tasks. Public reaction was both awe and fear, with many wondering what this meant for the future of AI.

Now, before the internet, data was like gold dust for AI researchers—scarce and hard to come by. But as the internet exploded, so too did the amount of data floating around. Suddenly, there were mountains of information generated by websites, emails, online transactions, and more.

With the internet, AI could tap into vast amounts of real-world data, making it possible to train more accurate and effective models. Moreover, the internet provided a new platform for AI applications. Take email, for example. Spam filters use AI to identify and block unwanted messages. These filters learned from massive datasets of emails, picking up on patterns and keywords that indicated spam.

The 2000s saw even more rapid advancements, largely thanks to the explosion of data and improvements in computing power. AI began to move beyond laboratories and into everyday life. For example, recommendation systems on platforms like Amazon and Netflix began suggesting products and shows based on user preferences. Search engines like Google started using AI algorithms to deliver more relevant search results. These algorithms analyzed user queries, clicked links, and browsing history to improve over time.

The real change, though, was the rise of machine learning and deep learning in the 2010s. These approaches, especially deep learning, which involves neural networks with many layers, allowed AI to make leaps in fields like image and speech recognition. In 2012, a neural network developed by researchers at the University of Toronto won the ImageNet competition, a prestigious image recognition contest, by a significant margin. This victory marked a turning point and demonstrated the power of deep learning.

From there, the progress was nothing short of extraordinary. AI started driving cars, recognizing faces, translating languages, and even beating humans at complex games. In 2016, Google DeepMind's AlphaGo defeated Lee Sedol, one of the world's top Go players, in a highly publicized match. Go is an ancient board game far more complex than chess, and AlphaGo's victory was seen as a major milestone.

AlphaGo used a combination of deep learning and reinforcement learning, which meant it learned by playing millions of games against itself and improving over time. It developed an ability to make creative and unexpected moves that even surprised experienced players.

The match was a best-of-five series. AlphaGo won the first three games, clinching the series. Sedol managed to win the fourth game, finding a chink in AlphaGo's armor, but AlphaGo bounced back to win the fifth game, making the final score 4-1.

The implications of AlphaGo's victory were profound. It wasn't just about winning a game, though that alone was surprising, much like Deep Blue's victory years before. However, it showed the potential of AI to tackle incredibly complex problems. The strategies AlphaGo used were studied by Go players worldwide, and the techniques developed for AlphaGo have been applied to other fields, like protein folding in biology and optimizing data centers.

Today, AI is in smartphones, helping us navigate with GPS and providing us with smart assistants like Siri and Google Assistant. It's in homes, with devices like smart thermostats and home security systems. It's "invading" various industries all over, from healthcare, where it assists in diagnosing diseases and personalizing treatments, to finance, where it detects fraud and manages investments.

AI has come a long way from its early days of theoretical exploration. It experienced plenty of ups and downs, but the progress is undeniable. Researchers are now working on advancing AI to even greater heights, going back to their original interest in areas like General AI, which could potentially perform any intellectual task a human can.

Let's look at some of these future possibilities, like quantum computing. Regular computers have gotten pretty powerful, but they are practically incomparable to quantum computing. Quantum computers use the principles of quantum mechanics to process information in ways that traditional computers can't even dream of. Imagine solving complex problems in seconds that would take today's fastest computers thousands of years.

So, how does this tie into AI? Quantum computing has the potential to supercharge AI by giving it unimaginable processing power. More processing power means faster training times for machine learning models and the ability to tackle problems currently out of reach. Perhaps you could see breakthroughs in drug discovery, optimizing logistics networks, and even cracking some of the toughest encryption algorithms.

AI and humans are getting better at playing nice together. AI-human interaction is an area ripe with innovation. One of the coolest developments is in natural language understanding. AI is getting better at understanding not just what we say but what we mean. It goes beyond basic commands to grasping context, emotions, and, yes, even your sarcastic remarks about its quirks. It could pick up on your mood and respond accordingly or hold a conversation so naturally that you forget you're talking to a machine.

Then there's the field of augmented reality (AR) and virtual reality (VR). AI is being used to create more immersive and interactive experiences. One day, you might wear AR glasses that provide real-time translations or superimpose directions onto the real world as you navigate a new city. In VR, AI can generate dynamic environments and characters that adapt to your actions, making gaming and simulations more procedural and unpredictable than ever.

Another exciting area is emotional AI, where machines learn to recognize and respond to human emotions, potentially great for fields like customer service, mental health, and even education. You might one day see an AI tutor who senses when you're struggling and adjusts the teaching style to help you better understand the material.

Lastly, there are brain-computer interfaces (BCIs). This might sound like sci-fi, but it's becoming more of a reality. Elon Musk's Neuralink is pioneering brain-computer interfaces (BCIs). There have even been successful, FDA-approved trials. BCIs enable direct communication between the brain and a computer. This could lead to amazing advancements for people with disabilities. They could control prosthetic limbs or communicate through thought alone. It also opens up possibilities for immersive gaming and new ways of interacting with technology that we can't even fully imagine yet.

Chapter Two

Exploration of AI Technologies

Artificial intelligence is no longer a sci-fi concept or academic pursuit—it's rapidly becoming a part of our everyday lives, whether we realize it or not. From the virtual assistants on our phones to the recommendation engines that serve up our favorite shows, AI is quietly working behind the scenes to make our digital experiences smarter, more personalized, and more efficient.

But what exactly are the core technologies that power this AI revolution? In this chapter, we'll explain some of the most important and innovative domains within artificial intelligence. We'll explore everything from natural language processing and machine learning to cutting-edge fields like computer vision, robotics, and generative AI.

Along the way, we'll unpack these technologies' fundamental concepts, real-world applications, and mind-bending potential. We'll also take an honest look at the challenges and ethical considerations that come with such powerful capabilities.

Natural Language Processing

You've probably interacted with natural language processing (NLP) technology many times, like when using a smart assistant (Siri, Alexa, Google Assistant, etc.). Or maybe you've noticed how good your email client is at detecting spam messages, at least most of the time, or your writing app's impressive grammar and spelling suggestions.

So, what exactly is NLP? In simple terms, it's a field of artificial intelligence that enables computers to understand, interpret, and generate human language. Then, it can comprehend your spoken or written commands and requests just like a human would instead of relying on rigid and extremely specific coding as a traditional program would.

NLP combines various computer science, linguistics, and machine learning techniques to process and analyze large amounts of natural language data. The data could be anything from books, articles, emails, social media posts, you name it. It has probably been used to train an AI model before. The process extracts meaningful insights, identifies patterns, and generates human-like responses when appropriate.

One of the key components of NLP is natural language understanding (NLU), which helps machines comprehend the meaning behind text or speech. It involves named entity recognition (identifying people, places, and organizations), sentiment analysis (determining the emotional tone), and intent classification (figuring out what the user wants to do).

For example, when you ask a virtual assistant, "What's the weather like in London today?" the NLU part of NLP breaks down your query, identifies the entities (London), the intent (checking weather), and other contextual clues to provide an appropriate response.

Another equally significant aspect of NLP is natural language generation (NLG), which lets systems produce human-readable text or speech, like generating emails, chat messages, news articles, creative stories, or code from simple prompts. You've probably seen examples of this with AI writing assistants or chatbots that can engage in surprisingly natural conversations.

NLP powers many applications you likely use daily: automatic text summarization, machine translation, sentiment analysis for customer feedback, or social media monitoring.

Of course, teaching machines to truly understand human language is an immense challenge. Humans communicate using complex structures, ambiguities, idioms, and contextual nuances that are second nature to us but can prove incredibly difficult for machines to grasp. That's why NLP systems rely on vast language datasets and sophisticated machine learning models to improve their capabilities continuously. But we already have AI assistants who can engage in witty banter or help us craft the next great novel. You just have to prompt them carefully and be mindful of their limitations. Language capabilities of AI, especially large language models (LLMs), have increased exponentially in only a few years.

Machine Learning and Predictive Analytics

The next major component of AI is machine learning or ML. Machine learning teaches computers to learn from data and make predictions or decisions without being explicitly programmed. Instead of following a set of rigid rules, machine learning algorithms use statistical techniques to identify patterns and relationships in massive datasets.

There are different types of machine learning, each with varying strengths and applications. Supervised learning, for example, involves feeding the algorithm labeled data and letting it learn to map inputs to desired outputs. This is how spam filters learn to recognize junk mail or how image recognition software learns to identify objects like plants or birds, even in crowded images.

On the other hand, unsupervised learning allows the algorithm to discover hidden patterns and structures in unlabeled data all by itself. This is useful for tasks like customer segmentation, where you want to group users by extrapolating similarities in their behavior or preferences.

Then there's reinforcement learning, which is like training a digital "agent" through trial and error and rewarding it for taking the right actions. This approach is used in areas like robotics, gaming AI, and even self-driving cars.

Machine learning powers many applications people rely on daily, often without even noticing. Voice assistants use machine learning for speech recognition and natural language processing. Online fraud detection systems also use machine learning to identify suspicious transaction patterns.

Of course, there is more. Machine learning is the enabler of predictive analytics, which involves using historical data to forecast future trends, behaviors, or events. For businesses, this can be incredibly convenient for optimizing operations, managing risks, and making decisions using cold, hard data.

If you need a more specific example of how it benefits businesses, let's imagine you're a retailer trying to plan inventory and staffing for the holiday season. Predictive analytics can help you analyze past sales data, customer behavior, economic indicators, and even weather patterns to forecast demand so you can have the right products in stock at the right time. Or let's say you're a healthcare provider looking for ways to improve patient outcomes. Predictive models could identify high-risk patients based on their medical history, lifestyle factors, and demographic data so you can intervene more proactively.

Predictive analytics is also used in fields as diverse as meteorology (for weather forecasting), sports (for predicting player performance and game outcomes), and criminology (for identifying potential criminal activity hotspots).

Of course, like any powerful technology, machine learning and predictive analytics come with some noteworthy challenges and considerations. One of the biggest issues is biases—if the training data is skewed or incomplete, the models can learn and perpetuate questionable biases and discriminatory patterns.

Privacy and ethical concerns are also a big topic of discussion, especially in sensitive applications like predictive policing or credit scoring. There's a tricky balance to strike between leveraging data for valuable insights and protecting individual privacy and civil liberties.

Then, there's the question of interpretability and trust. While machine learning models are becoming more complex (like deep neural networks), it can be challenging to understand how they arrive at their predictions or decisions. The lack of transparency can be a significant barrier to adoption in industries with strict regulations or high-stakes decisions.

But at the end of the day, machine learning is a powerful tool, but it is still a tool. It's up to humans to wield it responsibly, ethically, and with a clear understanding of its strengths and limitations.

Robotics and Automation

Robots are very much a part of the modern world, from factories and warehouses to hospitals and even homes. Fundamentally, robots are programmable physical machines that automate various tasks. They can take on a wide range of activities, from simple and repetitive actions to incredibly complex operations that require precision, speed, and endurance beyond human capabilities. However, the physical aspect of robotics sets it apart from AI. Of course, robots could be equipped with AI capabilities or connect to an AI model. However, they don't have to and can follow rigid instructions only.

One of the most common applications of robotics is in manufacturing and industrial automation. Robotic arms and computer-controlled machines have been a staple in assembly lines for decades. They have been building everything from cars and electronics to food and beverages. Their consistency, accuracy, and tireless nature make them indispensable for tasks that require repeating the same motions over and over again. But robots are no longer confined to factories. They're making their way into various other sectors, including healthcare, logistics, construction, and space exploration.

In hospitals, surgical robots assist doctors in performing delicate procedures with unprecedented precision, reducing complications and recovery times. Rehabilitation robots help patients regain mobility and strength through custom therapy sessions.

Autonomous mobile robots work alongside human workers in warehouses and fulfillment centers, transporting goods, managing inventory, and processing orders. With their help, human workers enjoy less physical strain while boosting efficiency.

On construction sites, robotic systems can perform tasks like bricklaying, tiling, and even 3D-printing of entire structures. These robots work tirelessly, even in hazardous conditions, and deliver consistent quality with minimal errors.

And let's not forget about space robots. From the rovers exploring Mars to robotic arms assisting astronauts on the International Space Station. These machines extend human reach and enable discoveries that would otherwise be impossible.

But these industrial applications aren't what's the most exciting. With advancements in artificial intelligence and machine learning, the rise of increasingly sophisticated service robots that interact with and assist humans in daily lives becomes imminent. Think about it—robotic vacuum cleaners and lawnmowers have already become commonplace in many households. Social robots with lifelike expressions and gestures are being developed to serve as companions, caregivers, and even educators for children. We're inching closer to a world where autonomous vehicles could take over transportation and logistics, potentially reducing accidents, congestion, and emissions.

Expect to see robots taking on even more complex roles, from exploring the depths of the oceans to assisting with disaster response efforts and possibly even serving as companions for the elderly or those with disabilities or mental health challenges.

But, of course, the rise of robotics and automation isn't without concerns. One of the biggest debates revolves around the potential impact on jobs and employment. Robots are becoming more capable of handling tasks traditionally performed by humans, so there are fears of widespread job displacement and economic disruption.

However, not to be discouraged, many experts argue that while some jobs may indeed be affected, robotics and automation will also create new ones, just as previous technological revolutions have done throughout history.

Another challenge is the safe and ethical development of robotic systems, particularly those that interact closely with humans or operate in public spaces. Rigorous testing, safety protocols, and guidelines are all necessary to mitigate risks and build public trust. Privacy and security are also critical considerations, especially as robots become more connected and capable of collecting and processing sensitive data.

Computer Vision

A major part of AI functionality that might surprise you is its ability to see. Think about how machines could see and interpret the world around them just like humans. That's the incredible potential of computer vision—a field of artificial intelligence that trains computers to extract meaningful information from digital images, videos, and other visual inputs, static or in motion. Computer vision combines principles from various disciplines, including machine learning, signal processing, geometry, and neurobiology, to mimic and enhance the human visual system's capabilities.

One of the most well-known applications of computer vision is in self-driving cars and advanced driver-assistance systems (ADAS). These systems can detect and recognize objects like pedestrians, traffic lights, lane markings, and other vehicles in real time using cameras, sensors, and sophisticated algorithms. This is what helps them drive safely and make split-second decisions about how to navigate a road challenge.

Then there's healthcare. Here, computer vision assists doctors when analyzing medical images like X-rays, CT scans, and MRI data with greater accuracy and efficiency. Computer vision can assist in early disease detection, treatment planning, and robotic-assisted surgeries through training algorithms to detect patterns and anomalies.

In security and surveillance, computer vision algorithms analyze footage from CCTV cameras to pinpoint potential threats, track people, and recognize facial features or license plates. This technology is already used in airports, public spaces, and law enforcement applications.

Computer vision is also used in retail, where it's powering advanced analytics and insights. By analyzing in-store camera footage, retailers track customer behavior, monitor inventory levels, and optimize store layouts and product placements. Some companies are even experimenting with computer vision systems that can detect and prevent shoplifting incidents.

The technology doesn't stop there. In agriculture, it monitors crop health and detects pests or diseases. In manufacturing, it's used for quality control and defect detection, and even in sports, it analyzes player movements and performances.

At the heart of computer vision are sophisticated machine learning algorithms trained on vast datasets of labeled images and videos. These algorithms learn to recognize patterns, identify objects, and extract valuable information from visual data.

One of the most popular techniques in computer vision is using convolutional neural networks (CNNs), a type of deep learning model inspired by the human brain's visual cortex. CNNs are incredibly effective at recognizing and classifying objects in images, which has made them the backbone of many computer vision applications.

Another aspect of computer vision is object detection and tracking, which involves identifying objects and then locating and tracking their movements across multiple frames or scenes. Autonomous vehicles, surveillance, and augmented reality stand to benefit the most since they deal with moving objects.

But, as always, there's more in the works. Image segmentation (separating foreground objects from backgrounds), 3D reconstruction (creating 3D models from 2D images), and facial recognition (identifying individuals from their facial features) are some of the hottest developments in computer vision.

Of course, computer vision has a few challenges and ethical considerations like any powerful technology, especially all those involved with AI. Privacy and data protection are major ones, especially in facial recognition or surveillance. There are also concerns about potential biases and inaccuracies in the algorithms, which could lead to unfair or discriminatory outcomes.

As we've seen already though, AI technology doesn't let such challenges hold it back. The field of computer vision is fueled by advancements in hardware, computing power, and algorithmic breakthroughs. We continue to generate more and more visual data through smartphones, drones, security cameras, and other sources, so the demand for effective computer vision solutions will only grow.

But beyond just practical applications, computer vision is also pushing the boundaries of what we thought possible in terms of machine perception and understanding. As these systems become more sophisticated, they may unlock new insights into how the human visual system works and perhaps inspire new approaches to artificial general intelligence (AGI).

Speech Recognition

Have you ever marveled at how seamlessly your smartphone's virtual assistant can understand your voice commands or how transcription software can accurately convert spoken words into text? That's the power of speech recognition technology at work. Speech recognition, also known as automatic speech recognition (ASR) or computer speech recognition, is the incredible ability of machines to process and understand human speech.

It's a technology that has been around for decades, but recent advancements in artificial intelligence, particularly in the fields of machine learning and natural language processing, have propelled speech recognition to new heights. This tech converts the complex analog audio signals of human speech (unknowable to the computer on a silicon level) into "ones and zeros"—the machine language computers can understand, analyze, and interpret. It usually happens in several stages, including signal processing, feature extraction, and pattern matching against pre-trained acoustic and language models.

One of the challenges you may commonly hear about in speech recognition is dealing with the vast variability in human speech. We all have different accents, speaking styles, and rates of speech, not to mention the impact of background noise, audio quality, and other environmental factors. To overcome these challenges, speech recognition systems now rely heavily on machine learning techniques, particularly deep learning models like recurrent neural networks (RNNs) and long short-term memory (LSTM) networks.

These advanced models are trained on massive datasets of transcribed speech samples, allowing them to learn and adapt to the intricate patterns of human speech. As they are exposed to more and more data, they become better at recognizing and transcribing speech accurately, even in tricky conditions.

Moreover, speech recognition can't stop at transcribing words—it's also about understanding the meaning and intent behind those words. This is where natural language processing (NLP) comes into the picture by combining techniques like intent classification and named entity recognition with speech recognition. That's how modern systems can transcribe speech, interpret the user's intentions, and extract relevant information, such as names, dates, locations, and more. So, combining speech recognition and NLP is what powers many of the virtual assistants you might be using daily.

This AI-powered technology also finds its way into business and various commercial and public sectors. For one, in the healthcare sector, speech recognition streamlines medical documentation, allowing healthcare professionals to dictate notes and reports that are automatically transcribed, saving time and reducing the risk of errors. Speech recognition also allows for more efficient and accurate transcription of courtroom proceedings, depositions, and other legal events, capturing information verbatim. Meanwhile, in education, speech recognition is used to develop language learning tools and assistive technologies for students with learning disabilities or speech impairments. In customer service, speech recognition powers many automated call center systems for efficient routing, self-service, and real-time transcription and analysis of customer interactions.

Speech recognition, which was already fairly impressive even before the advent of advanced AI, has come a long way. However, there's still room for improvement. Accuracy can still be an issue, particularly in noisy environments or with accents and dialects that are underrepresented in training data. Privacy and security are also concerns, as speech recognition systems collect and process sensitive audio data. However, the potential benefits of speech recognition technology are immense. It makes for more natural and efficient human-machine interactions, so it can enhance productivity, accessibility, and convenience across various domains. Moreover, as speech recognition becomes more integrated with other emerging technologies like augmented reality, virtual reality, and the Internet of Things (IoT), we may witness entirely new applications and use cases that we can barely imagine at this point.

Deep Learning

Deep learning is the cutting-edge technology driving many of the most impressive artificial intelligence breakthroughs in recent years. So, what exactly is it? Put simply, it's a subset of machine learning inspired by the structure and function of the human brain. It involves training artificial neural networks—computational models loosely mimicking the interconnected neurons in our brains—to learn from vast amounts of data in a way that resembles how humans learn and adapt.

However, what sets deep learning apart from traditional machine learning is the ability to automatically learn and extract complex features and patterns from raw data without relying on extensive manual feature engineering or domain-specific knowledge. This is where it's the closest to a human mind, capable of "thinking" for itself, loosely speaking. That allows deep learning models to tackle incredibly complex problems previously considered too challenging or impossible for machines to solve.

One of the main enablers of deep learning (and the big reasons why it hadn't been feasible decades prior) is the availability of massive datasets and the computing power required to train these intricate neural networks. Thanks to the explosion of digital data and advancements in GPU technology, we can now feed these models with millions or even billions of data points, allowing them to learn and generalize in ways that were simply unimaginable just a decade ago.

While there are various types of neural network architectures, one of the most popular and influential in deep learning is the convolutional neural network (CNN). By processing visual data through multiple layers of interconnected neurons, each detecting different features like edges, shapes, and patterns, CNNs learn to recognize and classify objects in images with remarkable accuracy, surpassing even human performance in many cases.

Another widely used deep learning architecture is the recurrent neural network (RNN), which is particularly effective for processing sequential data like text, speech, and time series data. RNNs have advanced natural language processing (NLP) applications, such as machine translation, text generation, and sentiment analysis.

Deep learning isn't just limited to these specific architectures. Researchers are exploring new and innovative models, such as generative adversarial networks (GANs), which can generate highly realistic synthetic data, and transformer models, which have revolutionized language understanding and generation tasks.

Beyond just recognizing patterns and making predictions, deep learning also enables machines to generate entirely new content, from realistic images and videos to creative writing and even music composition. This is where deep learning intersects with the field of generative AI, pushing the boundaries of what we thought machines were capable of.

However, it's worth being mindful of the fact that these complex models require vast amounts of data and computing resources, which can be costly and energy-intensive. There are also concerns around bias, interpretability, and the potential for deep learning models to amplify or perpetuate societal biases present in their training data. Additionally, deep learning models can be susceptible to adversarial attacks, where well-thought-out but malicious input data can trick the model into making incorrect predictions or decisions.

Generative AI

Now, consider all these AI systems that can create entirely new and original content—from photorealistic images and videos to compelling stories, music, and computer code. This is the explosive new reality of generative AI, one of the most exciting and rapidly evolving fields in artificial intelligence. It refers to a class of machine learning models and techniques capable of generating new data rather than just analyzing or making predictions from existing data. These models are trained on vast datasets of text, images, audio, or other forms of data, learning to identify patterns and relationships to generate novel and original outputs.

One of the most well-known examples of generative AI is text-to-image models like DALL-E, Stable Diffusion, and Midjourney. These systems can create highly detailed and realistic images simply from textual descriptions, unleashing a world of creative possibilities for artists, designers, and even non-professionals.

Language models like GPT-3, PaLM, and ChatGPT can generate human-like text on virtually any topic, from creative writing and code generation to answering questions and engaging in open-ended conversations. These language models are trained on massive corpora of text data, which is how they learn patterns and structures of human language. Once they understand context, meaning, and style, they can generate coherent, grammatically correct, nuanced, and creative text.

Beyond text and images, generative AI is also making waves in audio and music generation. Jukebox, MuseNet, and Riffusion, for example, create original music compositions, melodies, and lyrics. This is amazing for both hobbyists who just want to hear their dream music come to life without having to learn music themselves and professionals who want to speed up production. Even creatives in other fields, such as writers, use these tools to bring their stories to life for themselves and their audience.

However, generative AI also has numerous practical applications across various industries that aren't strictly creative. In healthcare, for instance, generative models create synthetic medical data, such as realistic CT scans or MRI images, which can be invaluable for training diagnostic models without compromising patient privacy. In product design and manufacturing, generative AI can explore and create novel designs, optimizing for specific constraints and requirements. This could lead to more innovative and efficient products, from car parts and aircraft components to consumer goods and electronics.

It's no surprise that generative AI is making its mark in software development. AI-assisted coding tools can generate code snippets, suggest optimizations, help with debugging, suggest entire new approaches to coding, and even generate simple, fully functional applications.

But as powerful as generative AI is, it's not without flaws and limitations. One of the biggest is the potential for these models to generate harmful content, spread misinformation, or, perhaps most concerningly, create deepfakes—highly realistic but synthetic media that can be used for nefarious purposes. The uncertainty about what is real and what is AI-generated certainly isn't a comforting one.

There are also questions about the ownership and intellectual property rights of AI-generated content, as well as concerns about the potential impact on human creators and creative industries. And, while generative AI models can produce remarkably human-like outputs, they still lack true understanding and reasoning capabilities. They are essentially pattern-matching machines skilled at mimicking and recombining existing data in novel ways but without the deeper contextual understanding and common sense that humans possess.

Emotion Recognition

Imagine a world where machines can read your emotions like an open book. No poker faces or trying to hide how you really feel would work here—these advanced systems would have you figured out. That might sound a little creepy at first, but it's not as unsettling as it might sound. Emotion recognition technology is like giving computers a sort of emotional intelligence, allowing them to pick up on all the subtle cues that reveal people's inner feelings and mental states.

The AI isn't detecting only what one might expect—the obvious big grins or frowns. These AI models analyze the tiniest micro-expressions, voice intonations, and physiological signals most humans might overlook. A raised eyebrow, a slight lip twitch, a change in skin conductance—to emotion recognition systems, it's all data that can be decoded to sense anger, joy, fear, you name it.

One of the most visible applications is facial coding and expression analysis. Maybe you've seen demos of apps that can map all the minute facial movements as you react to something. That's computer vision plus emotion AI at work. It's like having a robotic expert in reading micro-expressions but at a massive scale.

Potential uses range from geeky fun (amazing video filters that animate characters based on your facial expressions) to more profound (assisting therapists and counselors in better understanding their clients' emotions). Researchers are even exploring emotion recognition for lie detection, though it's worth taking those claims with a grain of salt for now.

But faces are just the start. These models can also decode the emotional undercurrents in the way humans speak and write. Think of how useful that could be for virtual assistants and chatbots. It could mean no more awkward, robotic responses when you feel frustrated or upset. With emotion recognition, they could adapt their personality and language to be more supportive and empathetic. Extending this to affective computing, we could see emotionally intelligent tutoring systems that sense when a student is bored or confused and adjust their teaching approach. Or there could be empathetic companion robots that provide a supportive presence for the elderly or those with disabilities.

The sci-fi visions get even more fascinating. Imagine an autonomous vehicle that could actually gauge the emotional state and intentions of pedestrians and other drivers. It could mean the end of road rage incidents or tragic accidents caused by miscommunication.

Now, as amazing as this all sounds, there are certainly reasons to be cautious about the rise of emotion recognition tech. Privacy is a huge one. Do we really want our devices and apps to map our deepest feelings all the time? There are problematic surveillance and psychological manipulation implications to consider. Then there's the question of bias and accuracy. Like many AI systems, emotion recognition is only as good as the data it's trained on. If that data is skewed or lacks diversity, the models could struggle with cross-cultural emotional intelligence.

There are also some deep philosophical questions surrounding whether machines can ever truly "understand" human emotion beyond just detecting physical signals. Emotions are complex, nuanced, and shaped by lived experiences machines can't replicate.

But despite the challenges, this technology is rapidly advancing, and it's easy to get excited about the incredible potential. Just picture more empathetic AI assistants, emotionally aware gaming experiences, or brighter possibilities in mental healthcare, all thanks to AI, which can understand us on a personal and emotional level. Who knows, maybe one day they'll even create an AI therapist who's a better listener than most humans. Hopefully, it'll still encourage us to open up to our fellow feeling beings. Emotions are one of the most beautiful human traits, after all—no amount of computing power can ever replicate the real, messy, deeply moving experience of being alive.

Emotion recognition technology, while impressive, brings up some unavoidable ethical issues, especially for businesses. Privacy and consent are at the top of the list because this tech involves analyzing personal data like facial expressions and voice tones. People must know how their data is used and give explicit permission. Without these safeguards, there's a risk of privacy invasion and unauthorized surveillance, which is undesirable both for the users and service providers.

Also, if the systems are trained on biased data, they can produce skewed results, leading to unfair treatment. It's easy to see how touchy this issue can be regarding emotional recognition. If we are to see AI-driven psychologists, teachers, comfort providers for the elderly, and so on, these tools must be proven to genuinely resonate with these users' feelings and provide provable improvement. "One size fits all" principles won't work here, nor will models that fail to consider factors like socio-economic status, culture, age, life experience, personal variables, and more.

Recommendation Systems

If you've ever found yourself lost in the vast digital jungle of Netflix, scrolling endlessly through movies and shows, only to stumble upon the perfect pick for movie night, you can thank the mighty powers of recommendation systems. These smart algorithms cut through the clutter and serve up personalized suggestions just for you. They use a variety of techniques to predict what items (movies, products, songs, articles, you name it) a particular user might enjoy or find relevant based on their past behavior, preferences, and other data points.

One of the most common and powerful approaches is collaborative filtering. This technique analyzes patterns of behavior across large user bases to identify people with similar interests and preferences. If you and a bunch of other users have overlapping tastes in movies, books, or music, the system will start recommending items that you haven't experienced yet but that crowd enjoyed. It's a bit like swapping favorite band recommendations with friends, except these algorithms can process millions of data points and connect the most niche, obscure dots you'd never thought to explore on your own.

Then, there are content-based filtering methods that analyze the attributes and metadata of the items themselves—plot summaries, genre tags, author styles, and more. For example, if the system knows you love witty romantic comedies with the most tangled cases of mistaken identity, it will ensure one of these finds itself on your feed soon enough.

These days, recommendation engines are usually hybrid systems that combine multiple techniques for maximum accuracy. They might use demographic data like your age or location to start with broad filters, then layer on behavioral signals from your viewing and purchasing history, social connections, and real-time contextual factors like time of day or device type. The most sophisticated versions incorporate advanced machine learning to continuously refine and optimize their recommendation models based on new data streams. Yes, those algorithms are studying your habits and trying to nudge you toward shows they think you are the most likely to binge-watch. It may seem a bit overbearing for some, but it's great in moderation and terrific for business. After all, customers will be coming back for more when they know they're shown exactly what they like.

While entertainment and e-commerce are perhaps the most visible use cases, recommendation systems are leveraged across countless domains. When you search on Google or YouTube, those ranked results are recommendations based on your history and behaviors. Post suggestions on social media feeds are decided by complex recommendation engines analyzing your interests and network.

Regarding workplace productivity, systems like Microsoft 365 Insights analyze your work patterns and digital trails to intelligently surface the most relevant documents, contacts, and resources for your current project or context. Uber's routing algorithms recommend ideal pickup points and routes based on traffic conditions and rider demand patterns. Even dating apps use recommendation systems to suggest potential matches based on compatibility scores from your profile details and behaviors within the app. So, in a sense, modern romance is brought to you by machine learning.

Of course, these systems aren't perfect. We've all experienced the frustration of irrelevant recommendations that make us wonder if the algorithms really know us at all. Accuracy is an ongoing challenge as the sheer scale of user data and potentially relevant items grows vaster.

Then, there are the gnarly ethical questions around privacy, manipulation, and bias amplification. If recommendation engines inadvertently prioritize popular mainstream content over niche interests, do they become self-reinforcing echoes of majority tastes? There are also concerns about systems influencing user behaviors in unsavory ways to maximize commercial metrics.

Chapter Three

AI's Role in Business

AI is everywhere, from the virtual assistants on your phone to the algorithms guiding your online shopping experiences. Businesses leverage AI to improve efficiency, enhance customer experiences, and drive innovation. Whether in retail, finance, healthcare, or manufacturing, AI is a major player in reshaping industries, creating new opportunities, and setting the stage for the future of business.

This section will show you real-world applications of AI in both daily life and business life, while also exploring how these two spheres intertwine.

AI in Everyday Life and Business

You might or might not be surprised to learn just how much artificial intelligence has already woven its way into our daily lives and interactions with businesses. From the moment you wake up to the sound of your

AI-powered smart speaker gently rousing you to the digital experiences that greet you throughout the day, AI is quietly working behind the scenes. In other words, this technology is all around you before your workday even begins and remains with you throughout your day.

Friendly virtual assistants like Siri, Alexa, or Google Assistant handle everything from setting reminders and alarms to queuing up your favorite playlists or podcasts. The speech recognition and natural language processing capabilities that power these AI helpers are constantly improving, making voice-based interactions more natural and intuitive every day.

AI is deeply embedded in countless apps, websites, and digital platforms with which many of us interact daily. Every time you search on Google, those finely tuned ranking algorithms are leveraging AI to sift through billions of web pages and serve up the most relevant results for your query. When you receive autofill suggestions in your email client or word processor, you can thank AI-powered language models that have learned patterns from massive datasets to predict your next word or phrase.

When marketing to users like you, companies are using AI for hyper-personalized targeting and dynamic creative optimization across digital ad campaigns. Behind that, oddly relevant promoted posts in your social feed are, unsurprisingly, more AI hard at work.

Naturally, AI is also playing an increasingly significant role in cybersecurity and fraud detection, whether you're on the customer or business side of things. AI models adapt faster than any human team to identify and thwart emerging threats as they continue to ingest and learn from vast troves of data on hacking attempts, malware signatures, and suspicious transaction patterns.

Of course, with each passing year, AI capabilities advance, and new potential use cases emerge across every conceivable industry. However, one common undercurrent unites them all: data. The machine learning models and algorithms that power modern AI are utterly reliant on the availability of large, high-quality datasets to train on. Whether it's catalogs of product information, reams of business documents, logs of customer interactions, or simply the trail of digital exhaust everyone generates through daily online activities, data is the linchpin enabling companies to push the boundaries of what AI can do.

AI's Significance in Modern Business

It's hard to overstate AI's profound and wide-ranging impacts. It represents a tectonic shift in how businesses can harness data to augment and automate decision-making across every facet of their operations, from deep analytical insights to intelligent process automation.

Take a manufacturing firm deploying AI-powered predictive maintenance. By ingesting sensor data and analyzing it against historical patterns, these systems can detect early signs of potential equipment failures and recommend preemptive actions—minimizing costly downtime and extending asset lifespans. Or consider how AI is transforming companies' marketing and customer engagement. Predictive analytics models can segment audiences with laser precision while AI-driven ad-targeting uncovers new pools of prospects ripe for personalized outreach.

For example, MarketMuse and Persado uncover new pools of prospects ripe for personalized outreach. At the same time, AI-powered chatbots and virtual assistants, such as Conversica and Drift, provide responsive, round-the-clock customer service at scale. Wrike and Monday.com are being used for project management to automate task assignments and predict project risks, thereby keeping projects on track. In marketing, companies like HubSpot and MarketMuse are leveraging AI for hyper-personalized targeting. The result is a major boost in revenue and loyalty.

Speaking of revenue, sales organizations are tapping into AI for everything, from lead scoring and forecasting to optimizing pricing strategies and deal terms based on historical transaction data. No more flying blind or relying solely on gut instinct when AI can light the way. The examples are practically endless across every industry. Finance and banking leverage machine learning for credit risk modeling, fraud detection, algorithmic trading, and so on. Retailers use computer vision to track inventory, analyze demand patterns, and enhance logistics. Media giants employ AI for content recommendation and dynamic ad optimization.

Online retailers like Amazon have fully embraced AI across their operations. Demand forecasting, supply chain optimization, pricing strategies, product recommendations, and even the robots whirring away in their warehouses. For instance, Llamasoft, ClearMetal, and FourKites use AI to track supply chains and optimize operations, including demand forecasting and logistics planning. Their customer service operations use AI for everything, from instantaneous language translation to sentiment analysis that routes frustrated callers to human support staff.

Stepping into the brick-and-mortar world, major retailers are adopting AI vision systems for functions like automated checkout, identifying stocked inventory levels, and analyzing security camera footage to study in-store traffic patterns and customer behaviors.

Businesses also integrate AI tools like Zendesk Answer Bot and Ada to boost customer support services with instant responses for enhanced customer satisfaction. Meanwhile, tools such as Ultimate.ai are automating routine queries across multiple channels.

The financial sector has arguably been one of the earliest and most prolific adopters of AI and machine learning. Credit card issuers use advanced fraud detection models to identify and flag potentially fraudulent transactions instantaneously. Banks employ AI for jobs like intelligent customer service chatbots to handle routine queries, automated loan assessment and underwriting, portfolio optimization, and so much more.

Even the stodgy, tradition-bound legal field is beginning to embrace AI for automating tedious document analysis, case research, and extracting key data points—freeing lawyers and paralegals to focus on higher-level tasks. As AI models become better at understanding and reasoning over complex legal language, we may see even more radical shifts in how legal services operate.

This technology is also catalyzing entirely new business models that would have been unthinkable just a decade ago or less. Just look at the seismic shifts AI has triggered in transportation and mobility with the advent of ride-sharing services and the autonomous vehicle revolution simmering on the horizon. Or how AI and machine learning have turbocharged innovation cycles for pharma companies developing novel drugs and personalized medicine solutions. You could argue that the data—where insights, predictions, and algorithms have become some of the world's most valuable commodities—owes its very existence to breakthroughs in AI and machine learning over recent decades.

However, while many are trying their hand at it, deliberate planning and iterative test-and-learn aren't always easy to accomplish. Data quality, governance, and ethical considerations are just a few of the hurdles companies must be mindful of as they adopt AI themselves. One of the biggest obstacles for many is actually an internal culture shift: truly embracing an AI-centric mindset that sees algorithms and predictive insights as core drivers of decision-making and strategy across the entire organization.

For example, a report by McKinsey & Company[1] found that companies leading in AI adoption have a culture that puts high stakes into data and AI literacy across all levels of the organization, contributing to better overall performance and competitive advantage.[2]

The migration from human-centric to machine-augmented processes and workflows requires companies to rewire how they recruit, train, and upskill their human talent. After all, in an AI-enabled enterprise, the human workforce doesn't just hand over the reins to the machines. They're meant to become partners in shaping and steering AI's potential.

Growing pains aside, one thing is abundantly clear: AI is an indispensable catalyst for innovation and a competitive differentiator in the business arena. Those companies that can deftly harness its power and fold it into their organizational DNA stand to gain significant advantages in operational scalability, speed-to-market, and continuous optimization.

Meanwhile, the businesses that dismiss or underestimate AI's significance risk are finding themselves swiftly outmaneuvered by more technologically progressive competitors—in the most extreme cases, rendered obsolete by the very forces of disruption they failed to embrace. Either AI will be a formidable tailwind propelling companies forward or a gale-force headwind perpetually pushing against their growth. There is no middle ground.

How to Leverage Different Types of AI

Each AI technique and technology brings its strengths and capabilities to the table. The key for savvy organizations is figuring out which tools work best for which jobs and then putting them to work in purposeful ways that drive meaningful impact across your operations.

Let's start with one of the most ubiquitous and accessible branches of AI: natural language processing (NLP). By training machine learning models on massive datasets of text data, NLP lets computers understand, interpret, and generate human-friendly language. It could unlock powerful use cases like customer service chatbots and virtual assistants that engage in free-form conversations while accurately capturing intents and queries in your company. Clients don't need to deal with frustrating touchtone phone menus, just natural discourse with an always-available AI agent. An example of this tech in action, SurferSEO, Frase, and Clearscope, use NLP to help create optimized content for higher search engine rankings.

NLP also enables intelligent document processing to automatically extract insights from contracts, reports, and other unstructured data. It powers real-time language translation for global organizations. They can even generate entire emails, articles, persuasive prose, and more from simple prompts.

Then, we have the broad field of computer vision and its ability to automatically analyze and derive insights from digital images, videos, and other visual inputs. For retailers, it means AI that can monitor store footage to track foot traffic patterns, gauge product interest, and automate checkout. Within manufacturing and industrial settings, computer vision makes predictive maintenance more feasible by continuously monitoring equipment through cameras to detect anomalies before failures happen. And across industries, these capabilities are useful for identity verification, physical security, and automating quality assurance inspections.

Moving into more specialized domains, we find AI subfields like speech recognition in voice interfaces and transcription services for physicians capturing patient notes hands-free. On the other side, AI-generated speech synthesis opens new frontiers in audio content creation and virtual assistants with more natural, expressive dialogue abilities.

The possibilities grow even more compelling when you combine multiple AI techniques. Take customer service as an example. An AI agent could ingest a chat transcript (from a chatbot) using NLP to understand intent and context, reference a knowledge base through semantic search, and then deliver a final spoken response via speech generation. That's just scratching the surface, though. When you fold in other AI disciplines like machine learning for predictive analytics, deep learning for autonomous systems, and generative AI models for creative applications, the enterprise opportunity space expands exponentially. A marketing team could use predictive and recommendation systems to hyper-personalize omnichannel experiences while generative AI creates hyper-targeted creative ads en masse. Within operations, deep learning algorithms optimize production planning, inventory management, and logistics routing in real-time based on fluctuating demand signals.

AI even extends into robotic process automation (RPA) for repetitive digital tasks and AI-powered physical robotics for flexible manufacturing and autonomous supply chains. The potential use cases are seemingly endless. What's remarkable is how quickly these powerful AI capabilities are evolving, becoming more accurate, robust, and accessible with each passing year.

Of course, getting the most value from all these different AI tools still requires careful planning, consultation, implementation, and continual refinement. It's not just layering on AI for AI's sake but intentionally mapping different AI techniques to your business challenges and desired outcomes. Identifying the right data sources, developing responsible AI governance frameworks, and strategizing for change management are just a few of the requirements for organizations to become true AI-driven enterprises. But for those who get it right, the rewards of harnessing this incredible technological toolbox will be transformative in imaginative ways.

The Importance of Data

All this sounds well and good for anyone running a company, but 'this wouldn't be possible without data. Without a constant stream of high-quality data to consume and learn from, even the most advanced AI would be hopelessly lost, like a supercomputer left scrolling endlessly through lines of useless code. Data provides the raw material that AI algorithms metabolize into insights, predictions, and intelligent decision-making capabilities. It's the informational lifeblood that allows machine learning models to continuously iterate, evolve, and sharpen their accuracy over time.

Training an AI is a bit like teaching a kid through intensive hands-on experience, flashcard drills, and an endless cycle of testing and feedback. Except in this case, the "experiences" are gargantuan datasets capturing every conceivable scenario, nuance, and edge case possible within that particular domain. The more quality data these AI students can crunch through, the smarter, more adaptable, and universally capable they become. It's one gigantic, never-ending lesson that hinges entirely on having the right data supplies at hand.

What this means for companies and organizations is that their ability to develop and leverage transformative AI is directly proportional to how effectively they can capture, manage, and refine the raw information flowing through their operations each day. Every customer transaction, supply chain data point, marketing interaction, and equipment sensor log are all little droplets in a vast ocean of big data that could fuel incredible AI breakthroughs if properly harnessed. Like any untamed resource, this data must be carefully extracted, cleaned, formatted, and refined into machine-readable formats before the AI systems can drink deeply.

Data governance, lineage, quality control, privacy, and security safeguards are just a few of the rigorous processes companies must master as they transition into AI-centric enterprises. After all, an AI model is only as intelligent as the data it's trained upon.

Imagine an AI assistant for customer support that can only comprehend queries about a narrow subset of products because that's the limited dataset its natural language processing model ingested during training. Or worse—a predictive model for medical diagnoses that produces skewed outputs because its data fails to represent certain demographics or health conditions. The constant thirst for high-fidelity data has made it one of the most precious resources. Unsurprisingly, the companies sitting atop vast data stockpiles have an incredible competitive advantage with AI. From FAANG tech giants like Google, Amazon, and Facebook, whose entire businesses generate torrents of user data, to vanguard enterprises like UPS, who crunch sensor logs across their global logistics network, the data-rich have a head start.

However, with the rise of big data platforms, data sharing ecosystems, and AI Cloud services like Microsoft, Google, and Amazon, any company can realistically tap into the AI revolution regardless of their data maturity level today. Much like electricity transformed every industry, AI capabilities are rapidly proliferating across the business landscape. Of course, simply having access to quality data isn't enough. Companies must develop robust data strategies that account for responsible AI governance, iron-clad data security and privacy controls, ethical data collection standards, and more. AI may be the future growth engine, but data will always be its nuclear fuel source.

Fortunately, as AI and machine learning continue advancing, these systems are also becoming more adept at processing messier, unstructured data formats. From images and audio to free-form text and raw video feeds, minimally refined data is increasingly fair game as training fodder.

So, while data remains the lifeblood of AI, the donors don't necessarily need to be perfect specimens. Diverse, comprehensive, and responsibly collected data is what AI systems truly thrive upon as they relentlessly extend their understanding of the infinitely complex world.

AI Across All Sectors

It's hard to find an industry that hasn't already been impacted by this technological force. Let's start with one of the most visible arenas where AI has taken center stage: retail and e-commerce. Those personalized product recommendations you see on Amazon are AI working behind the scenes, analyzing your browsing history and purchasing patterns to surface relevant items you're likely to buy. Have you ever wondered how retailers like Walmart can keep their shelves optimally stocked without constant human oversight? Computer vision systems that use AI to automatically scan inventories and forecast demand based on sales data deserve much of the credit. In fact, supply chain operations across industries like manufacturing, consumer goods, and logistics have all taken full advantage of AI and machine learning. Predictive analytics models help companies get ahead of potential snags by optimizing inventory flows, delivery routes, and production schedules based on reams of data.

Shifting over to the banking sector, AI safeguards customers by detecting fraudulent transactions as they take charge of loan approvals through automated underwriting systems. Some forward-thinking institutions are even using AI chatbots and voice assistants to provide personalized banking services and financial advice.

The power of AI to uncover insights from massive datasets has made it an indispensable tool within industries like healthcare and life sciences. Machine learning models can sift through medical images to assist radiologists in discovering tumors and other abnormalities. Predictive algorithms can map disease risk factors and assist in designing treatment regimens for individual patients. In fact, AI is at the forefront of personalized and precision medicine, where it accelerates drug discovery, optimizes dosing, and assists with robotic surgeries. The potential benefits in terms of patient outcomes and healthcare cost savings are staggering.

From curing diseases to crunching numbers, AI's disruptive influence stretches into finance and professional services. AI-powered equity analysis, automated trade execution systems, and quantitative risk modeling augment human analysts and advisors in new ways. If you've filed your taxes recently using software, you've likely encountered

AI-automated data extraction from financial documents and personalized advice tailored to your specific situation and deductions. As an example, Kabbage uses AI to automate lending decisions and provide small businesses with quick access to funding, while Kensho uses it to analyze financial markets and provide data-driven insights for investment. Then there is ZestFinance, an underwriting tool that helps lenders make more accurate credit decisions.

AI's wingspan also extends into realms like scientific research and engineering. AI systems are accelerating discoveries by processing vast datasets, simulating complex scenarios, and even generating novel molecular configurations for materials design.

However, one sector you might not immediately associate with cutting-edge AI is agriculture, but make no mistake, AI is cultivating a modern farming revolution. Computer vision lets autonomous robots and drones precisely identify diseases, pests, and soil conditions. Predictive modeling optimizes crop management and maximizes yields based on weather data and historical patterns.

Speaking of autonomous systems, AI is quite literally reshaping how we commute, travel, and even explore the frontiers of space. Self-driving cars, AI-powered logistics routing, and intelligent navigation systems are already enhancing mobility across industries like transportation, shipping, and aviation.

When it comes to industries tied to our planet's future, AI is also a conservationist and environmental guardian. Machine learning algorithms predict climate patterns, manage smart grids, and optimize renewable energy capture from solar, wind, and beyond. Conservation and sustainability efforts rely heavily on AI analysis to understand threats and interventions.

Of course, the business world's full-fledged embrace of AI goes far beyond individual use cases or industry applications. More fundamentally, AI is catalyzing new operating models, business processes, and entirely new product and service offerings companies never dreamed possible just a few years ago. AI-first startups are challenging entrenched incumbents who previously relied on legacy systems and human-centric workflows. While the impact of AI may manifest differently across sectors, one common thread unites them all: the ability to make smarter decisions from data-driven insights.

In an era of dizzying technological disruption and rapidly shifting consumer behavior, AI empowers organizations to be hypersensitive and adaptable—no longer forced to rely solely on historical precedents or human gut instincts. With AI's predictive powers augmenting their workforces, businesses across industries can stay ahead of the curve and capitalize on opportunities before their competitors even know they're there.

So whether through intelligent automation achieving new frontiers of productivity, forward-looking forecasts driving proactive planning and risk mitigation, or AI-powered services that wow customers with personalization and convenience, this transformative force has already woven itself into the very fabric of how we do business today.

1. McKinsey & Company. (2024). The state of AI in early 2024: Gen AI adoption spikes and starts to generate value. Retrieved from https://www.mckinsey.com/~/media/mckinsey/business%20functions/mckinsey%20analytics/our%20insights/the%20data%20driven%20enterprise%20of%202025/the-data-driven-enterprise-of-2025-final.pdf

2. McKinsey & Company. (2022). The state of AI in 2022—and a half decade in review. Retrieved from https://www.mckinsey.com/~/media/McKinsey/Business%20Functions/McKinsey%20Analytics/Our%20Insights/Why%20data%20culture%20matters/Why-data-culture-matters.ashx

CHAPTER FOUR

DEMYSTIFYING AI - SEPARATING FACT FROM FICTION

AI HAS BECOME A hot topic, sparking both excitement and fear. It's no surprise, given the incredible things AI can do and the dramatic stories we hear about its potential. But with all the attention, it's easy for myths and misconceptions to spread. In this chapter, we're taking a closer look at some of the most common myths about AI.

From fears of job-stealing robots to the idea that AI can surpass human intelligence or make perfectly unbiased decisions, we'll dig into what's real, what's pure fiction, and what might sit somewhere in between. We'll also explore the realities behind AI in business, tackling the myth that AI is only for large corporations. It's time to separate fact from fiction and get a clearer picture of what AI truly means for our future. It will remain a part of our reality for a long time, so it's clear why debunking myths early on matters so much.

The Nature of AI Myths

There's no shortage of myths, misconceptions, and outright fear-mongering surrounding this incredible but often misunderstood technology. From dire predictions of a Terminator-style robot apocalypse to exaggerated claims of AI becoming a hyper-intelligent overlord, separating fact from fiction can be a daunting task. It's understandable to feel this way when stepping into anything unknown, especially considering that some AI woes have real-life justifications. However, to say that not all is so black-and-white would be an understatement.

At the heart of many AI myths lies a fundamental lack of understanding about what this technology actually is and isn't capable of, both today and in the foreseeable future. Cue the mental images of humanoid robots exhibiting human-level consciousness and emotional intelligence straight out of a sci-fi film.

But the reality is AI systems, no matter how advanced, are still firmly confined to operating within narrow, highly specialized domains dictated by their training data and the problems they're designed to solve. True artificial general intelligence (AGI) with broad, human-like reasoning and cognition remains an elusive, hypothetical concept at the moment.

So, where do these pervasive AI myths come from? Much of the hype, fear, and futurism can be traced back to the technology's origins and societal impact over the decades. The very concept of artificial intelligence has been stirring imaginations (and anxieties) since the 1950s when computer scientists and mathematicians first began speculating about the possibility of creating "thinking machines." As theoretical AI research progressed into the digital age, each new milestone, like expert systems, neural networks, and eventually machine learning, triggered waves of public discourse and debate.

Will these systems eventually become self-aware? Could they somehow break free of their coded constraints and subjugate their human creators? The narratives practically write themselves for sci-fi authors and moviemakers to spin into dystopian tales of machine insurrection. What's more, AI's exponential progress and rapid real-world deployment over recent years have only amplified these existential musings. More practically, though, the looming societal impacts and unintended consequences of increasingly capable AI systems have also fostered valid ethical concerns.

What happens as AI automates certain jobs and disrupts entire sectors of the workforce? How can we prevent these systems from amplifying existing human biases? What are the privacy implications as AI systems become ubiquitous analytical aids? These are just a few of the thorny issues vexing policymakers and fueling heated public discourse. Unsurprisingly, then, the myths surrounding AI often stem from sensationalized doomsaying about job decimation or exaggerated, anthropomorphized characterizations of AI as a threatening, uncontrollable force. The sci-fi tropes of machines subsuming humanity run deep.

But look beyond the hype, and you'll find more substantive debates around complex ethical and philosophical issues like privacy, transparency, accountability, and the delineation of human-AI boundaries in sensitive decision-making processes. Myths become metaphors grounded in very real societal concerns.

From a purely technological capabilities perspective, there's ample room for myths to take root around what AI can and cannot actually do, given the incredible speed of advancement. Conflating narrowly specialized systems like medical diagnosis AIs with the idea of superintelligent AGI is a common faux pas. Or consider how language models like ChatGPT that can engage in stunningly coherent written dialogue might be misconstrued as having true sentience or general intelligence when, in reality, they're just incredibly sophisticated text predictors lacking deeper understanding and reasoning ability.

So, while AI's impact and potential across sectors like healthcare, education, scientific research, and governance is quite real, divorcing AI fact from fiction requires distinguishing between currently viable use cases and the aspirational long-term goals many mythmakers speculate about.

At the end of the day, many AI myths arise from very natural human tendencies. Our innate revulsion to the notion of ceding control and capabilities to non-human entities, our fears about potentially disruptive emerging technologies, and our tendencies to anthropomorphize and project human-like traits and intentions onto inanimate objects. But just because these deep-seated psychological origins exist doesn't make all AI anxiety and myth-making invalid or unworthy of thoughtful examination.

Myth 1: AI Will Lead to Massive Job Losses

One of the most pervasive and anxiety-inducing myths surrounding AI is the idea that it will lead to widespread job losses and economic devastation as machines inevitably replace human workers across most industries. You've probably seen the doomsday headlines proclaiming, "Robots Are Coming for Your Job!" Let's all take a deep breath and inject a healthy dose of reality into this admittedly unnerving (and certainly not entirely baseless) scenario. While AI will undoubtedly change the workforce and individual job roles in profound ways, jumping straight to visions of permanent mass unemployment is not only misguided but also grossly oversimplifies AI's nuanced relationship with human labor.

Let's start with some hard data. Despite the rapid acceleration of AI adoption across sectors over the past decade, job losses directly attributable to automation and AI have been modest so far. In fact, the World Economic Forum estimated that by 2025, AI could create nearly 60 million more jobs than it destroys.

Of course, this is little consolation if you happen to be among those displaced workers in administratively intensive roles like data entry, accounting, or customer service, the types of highly routine, repetitive jobs ripe for automation by robotic process automation and conversational AI assistants. But here's the unfortunate but undeniable truth: job displacement due to technological change is nothing new. From the Industrial Revolution to the dawn of the Internet age, innovation has always disrupted certain labor markets while simultaneously creating new employment avenues that were previously unimaginable.

Take a look at job categories like data scientist, machine learning engineer, AI ethicist, and robotics technician—these are just a few examples of lucrative, future-proof career paths that essentially didn't exist twenty years ago until AI started gaining traction. So, while AI will certainly automate some existing tasks and even entire roles, it will also give rise to entirely new categories of jobs that we can't yet foresee or comprehend today. That's the creative destruction cycle of technological progress in action.

And let's not forget—AI is, first and foremost, a productivity multiplier. By intelligently automating rote tasks, AI augments and empowers human workers, freeing them up to focus on higher-value activities that machines still struggle with, such as strategic planning, creativity, empathy, and complex analysis. Take healthcare as a prime example. While AI systems are already assisting radiologists in spotting anomalies and streamlining medical imaging, the technology is also enabling physicians to spend more quality face time with patients by offloading tedious data entry and paperwork burdens.

And consider the rise of the "gig economy" from AI and automation. Intelligent assistants, delivery drones, self-driving vehicles, and more open up all kinds of entrepreneurial opportunities for workers outside the traditional employment constraints through freelancing, contract work, and micro-tasking. Now, the gig economy itself has many potential issues, job insecurity and limited workers' rights being just some of them, but these issues run far deeper than AI and have been around way before its advent.

Of course, one can't ignore the very real pain and disruption that will happen as labor sectors contract or even disappear entirely due to increasingly capable AI automation. Truck driving, food service, and general labor are just some examples of industries that will face severe workforce pressures in the not-too-distant future. But instead of trying to put up artificial barriers, the imperative is to rethink workforce development, education, and retraining systems to funnel displaced workers into the new roles that AI is creating. Government policies, social safety nets, and public-private partnerships will be mandatory components of this transition.

This type of large-scale reskilling and labor mobility is no easy feat. However, the costs of not managing AI's impact on employment could be catastrophic—rising income inequality, social unrest, and widespread skills obsolescence, just to name a few consequences. A head-in-the-sand posture is not an option. So, while AI will indisputably cause labor dislocations, displacement and unemployment are far from the inevitabilities many fear. History has repeatedly shown that as old jobs become obsolete, new vocations eventually emerge.

The Wright brothers disrupted the entire equestrian industry, but it's hard to look back and say airplanes were a net negative for job creation and economic expansion in the long run. Something to ponder the next time you hear dire predictions about AI stealing everyone's jobs and bringing civilization to its knees. With the right workforce development policies and social safety nets in place, AI can and should be a catalyst for new employment prospects, entrepreneurial opportunities, and human productivity enhancements—not a looming job apocalypse to be feared and rejected outright.

Myth 2: AI Can Surpass Human Intelligence

The notion that AI will one day become so intelligent that it surpasses human-level cognition and capabilities is one of those classic AI myths that has been prompting both awe and existential dread since the earliest days of sci-fi storytelling. Just picture it: A future where hyper-advanced AI systems have transcended their original programs, exhibiting general intelligence, self-awareness, and even consciousness in ways that make us mere humans look utterly primitive by comparison. Machines are becoming the unequivocal dominant life form on Earth and beyond. That's some seriously unsettling yet admittedly compelling speculative fiction fuel.

But let's take a step back from dystopian Hollywood fantasies for a moment. While the idea of artificial superintelligence makes for gripping narrative premises, the reality is that AI is still incredibly far away from anything even remotely resembling generalized human-level brilliance and autonomy.

Today's AI systems, as impressive and capable as they may seem, are still fundamentally narrow and specialized tools confined to performing specific tasks within well-defined boundaries and contexts. Their "intelligence" is really just highly optimized pattern recognition and probability mapping based on their training data.

Sure, some of the latest language models can engage in stunningly coherent written dialogue almost indistinguishable from a human. But dig deeper, and you'll find these systems don't actually have true semantic understanding, reasoning abilities, or robust world knowledge. They're incredibly sophisticated parlor trick performers, not sentient beings.

In fact, once you start noticing prevalent "AI-isms," it's hard to unsee them. Indeed, the AI must first pick up on these phrases or modes of speech somewhere (meaning human-generated text) before it can repeat them. However, unlike a person, it can't understand when they're being overused. It may use certain phrases, metaphors, and so forth in ways that a human would know don't really work.

Even the most cutting-edge AI excels only within its specialized domain, whether parsing natural language, detecting anomalies in medical scans, or playing combinatorial strategy games at a superhuman level. Put those same focused intelligences in an unconstrained, open-ended environment and ask them to exhibit the type of general problem-solving, abstract thinking, and behavioral flexibility that even a human toddler can manage with ease. They fall hopelessly flat.

This is what's known as the AI "artificial general intelligence" grand challenge—endowing systems with the type of flexible, context-aware intelligence and broad reasoning that biological entities like humans take for granted. Many AI researchers believe this may be an intractable problem akin to trying to reproduce genuine self-awareness and consciousness rather than simply codifying narrow competencies.

Now, that's not to say the quest for AGI isn't an active area of serious research and development. AI pioneers are hard at work trying to crack the code of general intelligence, exploring computing architectures and paradigms like neural networks that more closely mimic how our brains process information. And once you bring ultra-advanced future technologies like brain-computer interfaces, quantum computing, or even mind uploading into the equation, then maybe, but only maybe, we could eventually blur the line between biological and artificial intelligence.

But based on the current technology trajectory, we're still likely many decades, if not centuries, away from seeing AI systems exhibit genuine autonomy, self-determination, and generalized intellectual prowess, eclipsing humanity's collective brilliance across every conceivable domain from the arts to the sciences. This brings us to another reason why the singularity-level "AI surpassing human intelligence" scenario seems far-fetched for the foreseeable future—the simple fact that humans are the ones developing and imbuing AI systems with their capabilities in the first place.

Just as parents teach children the fundamentals before they can become independent, AI systems rely entirely on the training data, architectures, and mathematical reward functions we painstakingly design for them. They have no ability to spontaneously reroute their core programming or transcend inherent constraints without explicit remodeling by their human developers. So while we may be creating increasingly sophisticated and specialized "genius machines" that trounce even the most skilled humans in games, pattern detection, or brute force computation, the

Promethean myth of AI one day spontaneously bootstrapping itself into an all-knowing, all-powerful superintelligence seems firmly grounded in science fiction rather than reality.

That's not to say we shouldn't carefully contemplate the ethical and safety implications as AI systems grow more potent and capable within their domains. Guidelines and failsafe measures must be in place to keep these powerful technologies safely constrained and aligned with human values as they solve increasingly complex problems.

But true human-level artificial general intelligence with the breadth and depth of skills to autonomously match or exceed our species' collective intellectual and creative prowess? That still remains much more within the realm of fable than imminent technological certainty based on the current state of AI. For now, at least, AI's real-world impacts are more about augmenting human ingenuity and abilities rather than directly superseding or replicating them entirely. The robot overlords can wait...probably.

Myth 3: AI Makes Unbiased Decisions

One other pervasive myth surrounding AI is the idea that these systems are inherently objective and unbiased decision-makers, free from the flaws and preconceptions that taint human judgment. After all, algorithms don't have egos, emotions, or ulterior motives clouding their analytical processes, right? They are just purely logical creations that coldly crunch the data and spit out the most rational, fact-based outcomes. But, as compelling as this "unburdened by human bias" narrative may sound, the reality is that AI models can actually reflect and even amplify biases in incredibly complicated ways.

The crux of the issue is that AI systems, for all their computational wizardry, are still entirely dependent on the data they're trained on to learn patterns and make decisions. If that training data contains historical human biases, sampling errors, missing information, opinions presented as truths, or skewed facts, the AI's outputs will perpetuate and compound those flawed perspectives at machine scale and speed.

Let's look at an example from the criminal justice system using AI technologies like predictive policing and risk assessment tools. If these systems are trained primarily on historical crime data from certain geographic areas or demographic groups that were already oversampled or unfairly scrutinized due to human biases, incorrect correlation, or circumstances at the time, then the AI will simply learn and institutionalize those exact distorted risk predictions moving forward.

Similar concerns have been raised about resume-screening AI tools that could inadvertently discriminate for or against job candidates if the training data exhibited even subtle human hiring biases against attributes unrelated to actual qualifications. Or consider AI smog-mapping and environmental justice scenarios where under-sampled data inputs fail to adequately account for low-income exposure risks.

Of course, AI bias doesn't just impact life-altering scenarios like the legal system or environmental protection. Product recommendation engines that only surface mainstream content could systematically underexpose niche creators. Automated loan underwriting models exhibiting bias could perpetuate unfair lending practices and wealth inequality. The list goes on across nearly every industry, and AI uses case imaginable, which is why most experts agree that far from being impartial arbiters, AI systems can actually concentrate and industrialize human prejudices if we're not extremely proactive about interrogating these models for skewed data inputs and adjusting accordingly.

For example, a study by the National Institute of Standards and Technology (NIST) emphasizes the need for addressing both human and systemic biases in AI systems, highlighting how these biases can be overlooked without careful consideration and management[1]. Moreover, McKinsey & Company discusses the complexities and challenges of fairness within AI models, bringing to light the need for different metrics and standards depending on the use case[2].

It also means having the humility to understand that AI is never a rigid panacea or absolute truth. It simply transforms the data we feed it into probabilities and suggestions. Human oversight, domain expertise, and cross-examination are still essential when scrutinizing AI model outputs and decisions, especially in high-stakes consequential use cases.

In other words, we can't just blindly treat AI as an inscrutable black box decision-maker, even if that's more convenient than subjecting its recommendations to further vetting and governance. Transparency in how an AI arrives at its outputs and clear protocols for human checks and accountability are paramount to preventing unfair or unintended outcomes at scale.

There's also the issue of bias from the AI development side—the teams actually defining the architectures, parameters, and objective functions we task these models to optimize toward. The good news is that there is more and more awareness across industries and the public sector of these critically essential challenges around ethical AI principles like fairness, accountability, and transparency through promising technical approaches like adversarial debiasing and discrimination-aware data mining.

Myth 4: AI Can Automate Complex Decision-Making

First things first, we need to understand what complex decision-making entails. We're not talking about simple "if this, then that" scenarios common to traditional programming here. The majority of traditional programs can be boiled down to and/or decision trees. But for complicated, real-world decision-making, this is far from enough. However, given AI's distinct capabilities, there's an expectation that it can make incredibly tricky decisions with utmost rationality. Complex decisions often involve multiple variables, ambiguity, and weighing trade-offs. Think about something like deciding on a corporate acquisition or determining the best course of treatment for a patient with multiple health conditions.

Now, AI has shown impressive capabilities in certain domains, like playing chess or Go, but these are ultimately games with well-defined rules and objectives. Real-world decision-making is a whole different ballgame. It's messy, nuanced, and riddled with gray areas that even humans struggle with.

AI systems have made significant strides in areas like medical imaging analysis and drug discovery. But when diagnosing and treating patients, things get a lot trickier. Every person's body is a little different, and a complex interplay of biological, psychological, and social factors shapes their circumstances. Even with all the data in the world, there's an irreducible element of human judgment and intuition necessary to tackle all this.

Ask any seasoned doctor about the times they had to make a gut call based on their years of experience, going against what the textbooks or algorithms suggested. That's not to say AI can't assist in the decisions.

It absolutely can and does by providing valuable insights and analysis. However, fully automating those high-stakes decisions is not an option, at least not with current technology.

Another area where complex decision-making rears its head is public policy. Imagine an AI system that determines immigration policies or resource allocation during a natural disaster. These decisions have far-reaching implications and involve weighing competing interests, ethical considerations, and long-term consequences. Even with the most advanced AI, there's simply too much nuance and too many intangibles for a machine to handle unaided.

Speaking of ethics, let's not forget that complex decisions often have a moral dimension. Like in the infamous trolly problem, is it ever okay to sacrifice a few for the greater good? These are the kinds of thorny questions that AI, for all its brilliance, struggles with. Without a proper understanding of human values and the ability to grapple with abstract concepts like fairness and justice, AI is ill-equipped to make these sorts of judgment calls on its own.

It's easy to imagine the counterarguments: "But what about AI systems that learn from massive datasets and mimic human decision-making patterns?" That is a fair point, but here's the issue: those systems are essentially approximating human judgment, not replacing it.

Myth 5: AI is Only for Large Corporations

Let's address the elephant in the room. Yes, some of the most cutting-edge AI developments have indeed come from the likes of Google, Microsoft, and other Silicon Valley giants. With vast resources and armies of researchers, they've been able to push the boundaries of what's possible with AI. But AI is no longer the exclusive domain of these behemoths. In fact, it's becoming accessible to businesses and individuals of all shapes and sizes. We're not just talking about simple chatbots or virtual assistants but powerful AI tools that significantly alter the way people work, create, and solve problems.

Take, for instance, small businesses. Traditionally, these scrappy underdogs have had to make do with limited resources and outdated technologies. With the rise of user-friendly AI platforms and services, even mom-and-pop shops can now tap into the power of machine learning and data analytics. You can have a local restaurant using AI to optimize inventory management and reduce food waste. Or a small e-commerce store using AI-powered product recommendations to boost sales and provide personalized shopping on par with the big players.

More intriguingly, independent creators, artists, and hobbyists are also getting in on the action. Thanks to tools like AI-assisted image and music generation, anyone with a computer and an internet connection can unleash their creativity in ways once unimaginable.

Speaking of creativity, let's talk more about how AI is shaking up content creation. Professional writers, designers, and marketers no longer have to toil away for hours, agonizing over every word and pixel. With AI writing assistants and design tools, anyone can produce high-quality content in a fraction of the time. It isn't just churning out generic, cookie-cutter content anymore, either. With the right prompting, the latest AI systems can actually adapt to your writing style, your brand voice, and your preferences so the content feels authentically yours.

Moreover, with AI-powered simulations and data analysis tools, independent researchers and citizen scientists can now tackle challenges that were once the exclusive domain of well-funded labs and universities.

You might even one day try to explore the mysteries of the universe or develop new materials. AI is giving the potential power to anyone with a curious mind and a thirst for knowledge to make groundbreaking discoveries.

And let's not forget about the role AI is playing in education and personal development. With AI-powered tutoring systems and personalized learning platforms, anyone can access high-quality education and upskill themselves, regardless of their location or financial circumstances. AI-powered translation tools break down language barriers, while AI-assistive technologies help people with disabilities live and learn much more independently.

---

1. National Institute of Standards and Technology. (2022). Towards a standard for identifying and managing bias in artificial intelligence (NIST Special Publication 1270). Retrieved from https://nvlpubs.nist.gov/nistpubs/SpecialPublications/NIST.SP.1270.pdf
2. McKinsey & Company. (2022). Tackling bias in artificial intelligence (and in humans). Retrieved from https://www.mckinsey.com/business-functions/mckinsey-analytics/our-insights/tackling-bias-in-artificial-intelligence-and-in-humans

Part II

Practical Applications and Case Studies

Chapter Five

AI in Customer Service

Imagine reaching out for help and getting an instant, accurate response without waiting on hold or navigating endless menus. That's the future that AI may promise you. Making customer interactions smoother and more efficient. In this chapter, we'll explore how AI, through chatbots and virtual assistants, is enhancing customer experiences by offering personalized service and proactive support.

We'll look at how businesses are using AI to understand and anticipate customer needs, turning feedback into actionable insights, and continuously improving their services.

Enhancing Customer Experience with Chatbots and Virtual Assistants

Imagine you're trying to book a flight online, but you hit a roadblock with the payment process. Instead of wading through pages of FAQs or waiting on hold for an eternity, you simply type out the issue in a chat window. Within seconds, a friendly virtual assistant pops up with a greeting and walks you through the solution step-by-step. Welcome to the world of AI-powered customer service. Chatbots and virtual assistants—AI systems that mimic human conversation and provide instant support are not just for simple tasks like checking order statuses or resetting passwords. The latest generation of AI helpers can handle everything from complex troubleshooting to personalized product recommendations.

Take Amtrak, for example. The railway company's virtual travel assistant, Julie, can do far more than just look up schedules. Powered by natural language processing (NLP) and machine learning, Julie can sift through Amtrak's vast database to answer detailed questions about baggage policies and station amenities and suggest the best routes based on a customer's preferences.

But that's just the tip of the iceberg. For example, Nvidia is taking AI assistants to dizzying new heights with powerful language models. Their virtual agent can engage in freeform dialogue, understand contextual nuances, and explain complex technical concepts in simple terms—like having a knowledgeable, patient expert on call 24/7.

And let's not forget about e-commerce giants like Amazon and eBay, which have integrated AI assistants directly into their platforms. With a few taps or voice commands, shoppers get product recommendations, track orders, and initiate returns, all without the need for human intervention, unless requested.

Now, you might be thinking, "Sure, AI assistants are great for simple queries, but what about really complex issues that require a human touch?" Fair question, but here's the thing: AI is getting scarily good at handling intricate, multi-layered conversations. Take IPsoft's Amelia. This AI can literally take over entire business processes, like managing insurance claims or onboarding new employees. Amelia can parse through massive amounts of data, decipher context and intent, and even escalate issues to human experts when needed. In other words, she's a super-intelligent digital colleague who never takes a coffee break.

However, perhaps the biggest game-changer is the true omnichannel customer experience. Customers expect smooth-sailing service across every touchpoint—website, mobile app, social media. You name it. AI makes that possible by creating a unified, 360-degree view of each customer's interactions and preferences.

Imagine being able to pick up a conversation with a virtual assistant right where you left off, whether you're on your laptop, smartphone, or a smart speaker, if that suits you better. You don't have to repeat yourself or feel like you're being passed from one siloed department to another. It's a level of personalized, contextual service that was practically unimaginable just a few years ago.

Naturally, however, as with any disruptive technology, there are valid concerns about the use of AI in customer service, such as privacy and data security. These systems usually have access to sensitive customer information, which no one would like to see leaked, apart from bad actors. There are also fears about job displacement, with some experts predicting that AI could automate up to 30 percent percent of customer service roles by 2030.

AI isn't meant to replace human agents entirely—it won't be easy to top the human touch any time soon. Instead, it augments the customer experience in ways that simply weren't possible without it by offloading routine tasks. Human agents can focus on more complex, high-value interactions that require empathy, creativity, and good old-fashioned emotional intelligence. As impressive as AI has become, there are still plenty of situations where a human touch is irreplaceable. Dealing with sensitive issues, resolving escalated complaints, or just providing reassuring personal connections—these are areas where human agents will reign supreme for the foreseeable future.

Personalization and AI

Have you ever felt like a faceless number in a sea of customers? Many have been there, and many more likely will be. Being bombarded with generic, impersonal marketing messages or treated like just another transaction doesn't make for an enjoyable experience.

Yet, thanks to the magic of AI, those days of one-size-fits-all customer experiences may eventually become a true relic of the past. Hyper-personalization is the new norm, where businesses customize products, services, and interactions to each and every customer's preferences, behaviors, and desires.

And no, we're not just talking about throwing your name into an email subject line and calling it a day. AI is enabling companies to personalize at a granular level that was practically unimaginable just a few years ago.

Take Netflix, for example. The streaming giant's recommendation engine is a masterclass in AI personalization. The algorithms analyze your viewing history, ratings, and subtle behavioral data, like how quickly you binge a new show, to predict your tastes with frightening accuracy. Even more interestingly, Netflix's AI can personalize the artwork, trailers, and the order in which content is presented to each user, all based on their individual preferences and viewing patterns.

And it's not just Netflix mastering the personalization game. E-commerce behemoths like Amazon and Alibaba have been using AI algorithms for years to deliver product recommendations and personalized search results, and they are dynamically adjusting pricing based on individual customer behavior.

AI recommendation systems are also making it possible to personalize the entire customer experience, from initial touchpoints all the way through to post-purchase support and retention efforts. Take Stitch Fix, the online personal styling service. Their AI algorithms analyze all kinds of data—style preferences and body measurements to the latest fashion trends and inventory levels—to hand-pick a personal box of clothing just for you. Then, when you provide feedback on each item, the AI gets smarter and more attuned to your tastes.

Now, consider how telecom giants like Verizon and AT&T use AI to personalize customer service and support. The system analyzes a customer's account history, usage patterns, feedback, and sentiment data from past interactions to identify potential issues or upsell opportunities.

However, perhaps the most mind-blowing examples of AI-powered personalization are in healthcare. Companies like Oneview Healthcare are using AI to customize in-room entertainment, educational content, meal plans, and therapy regimens using each individual's health profile and preferences. AI also helps healthcare providers personalize clinical decision-making and care delivery. Vast troves of medical data let AI systems identify risk factors and predict adverse events. They can then suggest optimal treatment plans that match each patient's genetic makeup and medical history.

When done right, with proper governance and ethical safeguards, AI-powered personalization has the potential to create immense value for both businesses and customers alike. For businesses, personalization drives increased customer loyalty and lifetime value, along with more efficient operations and targeted product development. And for customers, personalization can mean more relevant deliverables that cut through the noise and make them feel seen and valued as individuals. Most importantly, AI is making it possible to deliver white-glove, concierge-level service and support to the masses.

Predictive Customer Service

Here is something deeply interconnected with the aforementioned recommendation systems. Remember the days when you had to wait on hold for what felt like an eternity, only to get transferred from department to department, repeating your issue over and over again? Or when you had to sift through pages of unhelpful FAQs, never quite finding the answer you needed? Thanks to AI and predictive analytics, those frustrating experiences are quickly becoming a thing of the past.

At the heart of this are AI systems that sift through vast amounts of customer data, from past interactions and purchase histories to sentiment analysis and even external data sources like social media, to find patterns and predict future needs or issues. For example, Airbnb's AI analyzes a guest's past booking preferences, travel patterns, ratings they've left, and real-time data like weather and local events to suggest listings or experiences they might enjoy on their next trip.

As always, there's so much more to this. By analyzing historical data on common issues, pain points, and service inquiries, AI can proactively identify customers who are likely to run into problems and take preemptive action to address them. You might get a notification from your cable provider's virtual assistant saying, "We noticed there's been some internet outages in your area lately. Here are some troubleshooting steps you can try. Would you like us to schedule a technician visit just in case?"

AI can also predict when customers might be open to opportunities or at risk of turning away so companies can take steps to retain and nurture those valuable relationships. Think about how monumental that could be for telecommunications or subscription services. AI predicts churn risks so companies can intervene with promotions, personalized outreach, or incentives for each customer.

Perhaps the most exciting application of predictive customer service is in proactive problem-solving and support. Imagine if your car's AI assistant could detect a potential issue with your brakes based on sensor data and driving patterns and then proactively schedule a service appointment for you at your preferred dealership before it becomes a major problem. Perhaps your smart home appliances can predict when they might need maintenance or repairs based on usage patterns and historical data and then automatically order the necessary parts or schedule a service call for you without any intervention needed on your end.

These might sound like far-fetched scenarios, but companies like Mercedes-Benz, Samsung, and Bosch are already using predictive analytics to make proactive, predictive service a reality for their customers.

Instead of reactive, transactional engagements focused solely on resolving immediate issues, predictive service enables a more proactive, holistic approach that prioritizes anticipating and addressing customer issues and wants before they even arise.

Enhancing Email and Social Media Interactions with AI

Think back to any time you might've felt like you were screaming into the void when reaching out to a company's customer support channels. You know the drill: sending off an email or social media message, only to be met with deafening silence or a painfully generic, copy-pasted response weeks later. AI changes that, too. How do they accomplish that? They combine natural language processing (NLP) with machine learning and deep learning capabilities to actually understand the context, intent, and sentiment behind customer communications, enabling them to provide relevant responses at lightning speed.

Take Gmail's AI-powered smart reply suggestions, for instance. You've likely experienced the convenience of having the AI offer up contextually relevant responses based on the email thread. Imagine that same capability but turbo-charged and integrated directly into a company's customer service workflows. That's exactly what companies like Zendesk and

Salesforce are doing with their AI-powered email and social media automation tools. By training their AI models on massive datasets of past customer interactions, product information, and support documentation, they can intelligently respond to common inquiries, freeing up human agents to focus on more complex issues. It's not just canned, one-size-fits-all responses here. The latest generation of AI language models engage in multi-turn conversations, understanding context to provide dynamically generated, hyper-personalized responses.

Imagine reaching out to a company's support team on Twitter with a specific issue related to a recent purchase. Instead of getting a generic, unhelpful autoreply, the AI assistant actually ingests the details of your inquiry, cross-references it with your account information and purchase history, and provides a detailed, personalized response—all in real-time. AI lets companies proactively monitor social media channels and online communities for potential issues, complaints, or even just general sentiment about their brand. By quickly identifying and addressing these concerns before they spiral out of control, businesses can get ahead of potential PR nightmares and turn dissatisfied customers into raving fans.

Companies like Airbnb and Uber have been using this capability for years to scan social media for mentions of their brand and services and then engage with customers to address issues or even just offer a human touch during their experiences. One other pretty exciting aspect of AI-enhanced email and social media interactions is the ability to provide true omnichannel support that seamlessly bridges the digital and the physical world.

You could start a conversation with a brand's AI assistant on Twitter and then pick up that same thread via email, chatbot, or even in person at a physical store or service center without having to repeat yourself or lose context. Thanks to AI's ability to create a unified view of each customer's journey across every touchpoint, that level of continuity is quickly becoming a reality.

Levi's, the iconic retailer, has implemented an AI-powered virtual concierge that can assist shoppers across multiple channels, from their e-commerce site and mobile app to in-store kiosks and even voice assistants like Alexa or Google Home. By understanding each individual's preferences, purchase history, and real-time inquiries, Levi's AI can support its customers no matter where or how the customer chooses to engage.

This technology means increased efficiency, cost savings, and the ability to provide always-on, real-time support at an unprecedented scale for companies. But perhaps more importantly, it enables a level of personalization, continuity, and genuine customer understanding that simply wasn't possible with traditional, siloed support channels.

Sentiment Analysis

Imagine you're the head of customer experience at a major airline. There are massive amounts of data pouring in from every conceivable channel—emails, social media, call transcripts, surveys, you name it. But amid all that raw data, how can you gauge how your customers feel? What's driving their satisfaction or frustration? The name of the game is sentiment analysis, the AI secret weapon that's helping businesses like yours cut through the noise and zero in on the emotions, attitudes, and opinions that truly matter. It uses natural language processing (NLP) and machine learning to analyze human language and identify the underlying emotional tone—positive, negative, or neutral. Rather than doing simple keyword matching, Sentiment analysis can detect emotions like sarcasm, frustration, and subtle linguistic cues that would be virtually impossible for the human eye or ear to pick up on.

The ride-sharing giant Uber processes millions of customer support interactions every year across a dizzying array of channels. All that data is fed into an AI-powered sentiment analysis engine to rapidly identify potential hot spots of customer dissatisfaction and address issues before they escalate. But imagine if Uber's AI could detect a flurry of positive sentiment around a specific driver or geographic area and then use those insights to adjust marketing, promotions, or service offerings in those regions. Or picture being able to identify pockets of customers who are feeling particularly enthusiastic about a new product or service and then double down on targeting that segment with personalized offers or loyalty programs.

The applications are endless, and companies across virtually every industry are tapping into the power of sentiment analysis to gain a deeper understanding of their customers' emotions. Take the healthcare space, for example. By analyzing the sentiment behind patient feedback, reviews, and even clinical notes, healthcare providers can identify areas where they're excelling or falling short when it comes to delivering compassionate, patient-centric care. These insights can then inform everything from staff training and process improvements to facility design and patient experience initiatives.

Voice of Customer (VoC) programs and brand monitoring are also something to get pretty excited about. The idea is to be able to scour the depths of social media, online reviews, forums, and other digital channels to get a real-time pulse on how your brand is perceived and talked about in the wild.

That's exactly what companies like Salesforce and HubSpot are enabling with their AI-powered sentiment analysis tools. By training their models on massive datasets of online conversations and reviews, these platforms can rapidly identify positive or negative sentiment toward a brand, product, or campaign and even surface granular insights into the specific drivers behind that sentiment. Suddenly, you're not just flying blind based on anecdotal evidence or small sample sizes. Instead, you have hard data and actionable insights into what's truly resonating with your customers at scale.

And we're not just talking about analyzing text-based data here, either. The latest sentiment analysis models can also process audio, video, and facial expressions to gauge emotional reactions and engagement levels. Consider the potential of a call center's AI that could automatically analyze the sentiment behind customer phone interactions, identifying frustration or confusion in real-time and prompting agents with scripts or recommendations to course-correct on the fly. Also, consider how retailers could measure shoppers' emotional responses to different product displays or messaging through in-store cameras and facial recognition tech.

Of course, as with any technology that deals with personal data and emotional insights, there are valid concerns around privacy, ethics, and potential misuse that need addressing with care. These issues become even more pronounced with sentiment analysis due to its ability to infer emotions and opinions from various data sources. Privacy is a significant concern because sentiment analysis often involves processing sensitive information from social media, customer reviews, and other personal communications. Users must know how their data will be used, and companies should ask for explicit consent. This is also why data protection laws exist, such as the General Data Protection Regulation (GDPR) in

Europe and the California Consumer Privacy Act (CCPA) in the United States, to protect user data without holding back business.

Ethically, there is potential misuse if sentiment analysis is used to manipulate consumer behavior without their knowledge. So, when companies use sentiment analysis, they must respect people's autonomy and avoid exploitation for personal gain or profit, intentional or unintentional. As for the legal angle, businesses must adhere to regulations that govern data privacy and AI usage to comply with emerging regulations and mitigate risks.

But when implemented thoughtfully and transparently, with robust governance and consumer consent frameworks, sentiment analysis can be incredibly powerful for enhancing customer understanding and experiences. It can unlock a wealth of valuable insights that inform product development, marketing strategies, operational improvements, and service delivery models. By truly understanding the "why" behind customer behaviors and emotions, companies can get ahead of issues, double down on what's working, and foster deeper, more meaningful relationships with the people they serve.

As for the customers, sentiment analysis could create more personalized, empathetic, and intuitive interactions with those businesses. You should never have to repeat your frustrations or constantly provide the same feedback because the companies you deal with are already tapped into your emotional state and proactively addressing your needs. It's a level of emotional intelligence and customer intimacy that simply wasn't possible at scale before the advent of AI and sentiment analysis capabilities.

Perhaps most profoundly, sentiment analysis gives businesses a lens into the true voice of their customers—the authentic emotions, opinions, and attitudes that often get lost amid the deluge of structured data and metrics that have traditionally driven decision-making. In that sense, AI is more than augmenting human capabilities here. It's helping us reconnect with the core emotional drivers that have underpinned successful businesses since the dawn of commerce itself.

Using AI for Customer Feedback Collection and Analysis

You know that feeling when you take the time out of your busy schedule to fill out a customer survey or feedback form, especially if your feedback includes complaints about issues you've experienced or improvement suggestions, only to wonder if your input will make any difference. Those days may finally be over, thanks to AI. Companies are tapping into incredibly powerful NLP models to automatically analyze all that valuable customer feedback at scale. Let's say your business gets thousands of support tickets, social media comments, app store reviews, and survey responses flooding in every single day. Having human teams trying to make sense of all that unstructured data is like drinking from a firehose. But feed all that text into state-of-the-art AI language models, and suddenly, you can automatically extract insights, categorize comments, prioritize issues to address, and more.

One huge area where this shines is analyzing all the open-ended feedback fields in customer satisfaction surveys For a human analyst, combing through thousands of those for recurring themes and sentiments is like looking for a needle in a haystack. But NLP systems can parse through them all in seconds, accurately categorizing each one by topic, rating the emotional sentiment as positive or negative, and even pulling out specific mentions of product issues or feature requests.

Think about how powerful that is for companies trying to improve products and services. Instead of looking at the basic numeric scores, they can instantly drill down into all the qualitative feedback at a granular level. They can see exactly which product features delight customers and which ones are total duds that are not worth further investment. The AI can automatically surface all the constructive suggestions people are offering rather than having those get lost in the noise.

Indeed, companies could hire armies of analysts to manually read through and categorize all that. But that is extremely expensive, not to mention incredibly tedious work prone to human error and inconsistency. Having an AI assistant that can automatically handle that first pass is a massive improvement in terms of scalability and insights. For example, Consumer Reports has deployed AI tools to automatically read and categorize the millions of consumer reviews they get each year into product-level strengths, weaknesses, and use cases. It helps their human analysts quickly see the common threads across that deluge of data.

Of course, this goes way beyond surveys and reviews. AI language models can be set loose on all kinds of customer communications—support emails, chat logs, social media comments, and so much more. Using techniques like topic modeling, the AI can automatically cluster all the inbound feedback into recurring themes and frequently raised issues. So, a telecom company might immediately see that a huge percentage of their customers complain about inconsistent broadband speeds or spotty service in certain areas based on zip code data. Instead of digging through it all manually, the AI turns that huge pile of unstructured data into clear, actionable insights their teams can jump on.

The same goes for social listening around a company's brand, too. Feed all the tweets, Reddit posts, YouTube comments, and Facebook discussions a company might get to produce into an AI engine, and it can automatically bucket all the sentiment and surface potential crises or reputation threats that might be bubbling up. For example, maybe the AI notices an unusual spike in angry posts about a recent software update that bricked someone's device. That's exactly the kind of smoldering issue companies need to get ahead of with an apology or make-good offer before it turns into a raging PR disaster.

And thanks to AI, none of those insights have to stay siloed away, either. Rather than having dedicated teams digging through support tickets or social media logs, companies can feed all their disparate data sources into a central AI system. So, the product teams can automatically see all the feedback about design flaws in their latest update. The social media teams can see common complaints about misleading ad campaigns. Customer support gets an early warning about issues leading to a surge of inquiries for which they can staff up. Marketing can automatically spot all the praise about their latest viral campaign to double down on. Breaking down those internal silos is crucial for companies trying to get a comprehensive view of how customers truly experience their brand.

It's worth pointing out, as always, that getting the most value out of these AI analysis capabilities requires some legwork, too. Companies have to carefully structure their data pipelines and prep that unstructured text data to flow smoothly into their language models. They need comprehensive model training and human oversight to make sure the AI is accurately picking up on the right hints and edge cases. Executive buy-in is also necessary since there's always some resistance to "replacing" human teams with fancy new AI systems.

However, for companies willing to walk that path, customer feedback mining using artificial intelligence represents a paradigm shift in how they might turn that data into actionable gold. Companies can finally squeeze maximum insights out of every customer comment, converting that unstructured feedback into clear product roadmaps, service improvements, and strategic moves to retain customers and stay ahead of emerging threats.

Machine Learning and Continuous Service Improvement

Thanks to AI, companies can now analyze real-time usage patterns and feedback to automatically identify areas for optimization and tweak their offerings on the fly. Think about it this way: Your typical software company might release a major new product update every six to twelve months after poring over all the user feedback, bugs, and issues that have piled up since the last release. But what if they could just keep getting better every single day in an endless cycle of micro-optimizations and personalized experiences? That's the true power of combining cloud services, constant user telemetry, and machine learning models that automatically surface insights and suggested improvements.

'Think about how streaming media services like Netflix, Spotify, and YouTube effectively use AI to delight customers. They're ingesting constant real-time data on what you're watching or listening to when you pause, rewind, or abandon something partway through and tying that to attributes like genres, moods, actors, and more. Their machine learning engines then analyze those trillions of behaviors across their entire user base to make highly effective personalized recommendations and improve core product experiences at the same time.

Let's say their models detect that a certain percentage of users tend to drop off during the first ten or so minutes of a certain show or movie. That could be a signal there's a pacing issue that needs adjusting, like tightening up the opening to better hook viewers from the start. Alternatively, if tons of people are replaying certain scenes over and over, maybe there's an epic monologue or action set piece that could use special highlighting through editing or surfacing via curated playlist clips. The possibilities get even more poignant when you combine multimodal signals. For example, their AI could correlate dialogue transcripts, video analysis of facial expressions and scene composition, and user behavior data to pinpoint exactly which emotional arcs in a story are landing and which fall flat.

Over in the enterprise software, companies are adopting machine learning-powered monitoring systems that continually study how users navigate through various workflows, screens, and features. The AI can detect points of friction where there's excessive rage clicking, unintended exit loops, or just general confusion based on inputs and behavior patterns compared to expected paths optimized for efficiency.

For example, if a disproportionate number of call center employees seem to frequently struggle with a specific task flow in the CRM software (such as updating customer information), jumping between multiple applications and copying/pasting data, there's probably an opportunity to streamline that process into a more intuitive workflow. The machine learning model picks up on that pattern at scale and flags it for closer review and potential UI redesign. Similar approaches are being used in applications like healthcare record keeping and tax prep software to study how subject matter experts naturally prefer to navigate different tasks.

That's just scratching the surface, though. Nowadays, AI systems are not only analyzing digital clickstream usage behaviors but also actual voice and chat communications that customer service reps have with people. So, an AI assistant could automatically listen in and flag potential areas of confusion where customers seem to get stuck or frustrated based on the dialog exchanges, verbal tones, and emotional signals from speech transcripts. Afterward, the system could automatically route those specific calls to human reviewers to quickly diagnose and improve the troubleshooting procedures moving forward.

This also extends beyond software into physical products and services. AI predictive maintenance models crunching equipment sensor data and field reports can automatically identify units with parts showing excessive wear and tear compared to historical benchmarks, prompting preemptive service calls. Connected devices like smart home appliances, TVs, thermostats, and automobiles are constantly beaming usage telemetry back to the cloud, where AI models analyze it for emerging issues and improvement opportunities around reliability, performance, and new features to add.

Say a manufacturer's AI notices an uptick in error logs from a certain batch of products with subtle process variances from the factory floor. The system could automatically investigate those deviations, isolate the root cause, and tweak the production parameters before any major defects start slipping through. For companies running complex logistics operations, machine learning-powered route optimization happens continuously based on constantly shifting conditions on the ground.

In online services and mobile apps, sophisticated AI systems monitor every user interaction right on the spot—from their first landing page to click flows, UI gestures, error stack traces, and more. The models extract critical funnel insights, like where people bounce before completing actions or repeatedly get tripped up. That kind of granular visibility is gold for product teams identifying frustrating UX pain points to remove and the most impactful areas to double down on improving.

These same AI systems can run constant A/B tests and controlled experiments to automatically validate whether proposed UX tweaks, content updates, or new features genuinely move the needle on key metrics. Having this ability massively accelerates innovation cycles by avoiding the traditional months-long batch process of designing changes, shipping a new release, and then waiting for qualitative user feedback to trickle in slowly.

Chapter Six

AI in Marketing and Sales

Picture this: personalized content crafted effortlessly, customer demands anticipated with uncanny accuracy, and optimal strategies for the furthest reaches. AI is making everything smarter and more efficient by generating engaging content and boosting SEO to transform customer interactions with chatbots.

To some, AI content creation tools may seem more like toys than serious additions to their workflow. However, almost anyone who has been involved with marketing and sales knows just how tedious and time-consuming it can be. AI takes that burden off human creators. Therefore, they can focus on coming up with better strategies and more engaging content rather than handcrafting each email, sales pitch, or promotional flyer.

AI-Driven Content Creation and Curation

Those who think AI's disruption of creative fields like art, writing, and media is some far-off futuristic scenario are already living in the past. Advanced language models and generative AI are quickly becoming indispensable for brainstorming ideas, producing first drafts at breakneck speeds, and more. Let's start with something relatively straightforward, like using AI writing assistants to overcome writer's block or explore different creative concepts. You could prime a language model like GPT and Claude with some initial prompts, topics, or sample writing, then have it riff off those starting points with a flurry of long-form prose, outlines, plot concepts, and more in seconds, almost like an AI co-pilot constantly spitballing ideas to expand your creative aperture.

They can also digest large volumes of information and synthesize it into readable summaries, reports, articles, and so on. So rather than having human writers waste time slogging through mountains of research material, the AI can churn through it all, extract key insights, and generate first drafts, combining those facts into cohesive narratives to refine and build upon.

You could feed the AI all the research reports and clinical data for a new pharmaceutical product, then have it automatically assemble authoritative medical marketing content, patient education resources, and more. You can also give it access to your proprietary company knowledge bases, competitor intel, and market trend data to produce sales enablement material at a fraction of the usual time and cost.

That same model of research ingestion followed by regeneration could apply across fields like journalism, education, entertainment, and more.

Academics could take advantage of the AI's ability to read and distill full books down into summaries or literature review drafts, augmenting human expertise. Media companies could use it to rapidly produce templated articles and video scripts based on breaking news data feeds. Big brands are already exploring how AI creatives could brainstorm bold new ad campaign concepts and storyboard initial execution plans, massively accelerating ideation cycles compared to traditional methods.

All that opens up some fascinating hybrid human-AI creative models where the machine does the heavy lifting on initial drafts and concepts, but humans provide strategic oversight, editing, and polish to shape those synthetic outputs into finished products tailored just right for their intended audiences.

Content creation is just one side of the coin, though. The other major disruptive AI opportunity is in content curation and personalized recommendation engines. We're all familiar with services like Netflix, Spotify, and Amazon, surfacing hyper-relevant content picks based on our behavior patterns. However, state-of-the-art recommendation systems have evolved far beyond just looking at simple user attributes and viewing histories.

Nowadays, these AI systems can ingest and cross-analyze huge catalogs of user-generated opinions, reviews, social media commentary, and other auxiliary data signals to help triangulate what specific types of content users will be most receptive to. For example, they could look at the affinity clusters and sentiment trends around a newly dropped movie or album, factoring in tangential influencer discussions to pick up on more subtle tastes and positioning opportunities for different niche interest groups. Consequently, they allow creators to come up with novel ideas to serve these niches just as they wish.

The same principles apply to news aggregators using AI to create feeds of articles, podcasts, and video segments tailored to your unique mix of topical interests and preferred editorial sources, voices, and styles. Over time, the AI learns to hone in on your unique combination of passion domains while steering clear of your hot-button issue areas or other turn-offs.

In B2B (business-to-business), forward-thinking enterprises are exploring how to use AI curation that can ingest all their vast internal data repositories—product manuals, field guides, sales collateral, training docs, you name it—and create personalized learning portals for employees based on their roles, skill gaps, and areas for professional development.

You could have a video player that dynamically tweaks editing styles, pacing, visual effects, background music, story arc intensities, and more as it picks up on your emotional reactions through a webcam feed and correlates those against its training data from billions of other viewer sessions.

Maybe the next scene gets a touch more frenetic if you seem to get disengaged from slower exposition portions. Or a dramatic monologue gets punched up to an eleven if the AI senses you delighted in an actor's emotive delivery. And this isn't even science fiction. Early versions of those types of responsive narrative experiences are already being prototyped by companies like Netflix, Disney, and others.

Looking forward, generative AI models that dynamically remix or compose imagery, audio, videos, and text on the fly will unlock even more malleable interactive realms. Imagine an AI story verse engine that can infinitely repackage narrative elements in bespoke Franken-stories tailored precisely to your tastes, or a music generator that gives you an infinite supply of music in your favorite genre. No two people will experience the exact same personalized journey. Or open-ended sandbox creative experiences where the AI acts like an infinitely adaptable improv partner, riffing with you in art, music, game worlds, or whatever sandbox you dream up.

Of course, legal and ethical boundaries around copyrights, bias mitigation, misinformation filters, and other safety guardrails will be critical as this technology matures and scales. There's no putting this generative AI genie fully back in the bottle, though. These systems are already beginning to augment and reshape every creative field imaginable, from Hollywood studios to Madison Avenue Mad Men to independent artists and authors forging brave new storytelling frontiers.

Apps like Replika serve as AI companions to combat loneliness, while robots like Lovot, The Lovable Robot, and Paro Therapeutic Robot provide companionship and socialization. These bots may make users feel less lonely and enhance social skills, but they also raise ethical and psychological questions. There's concern that people might become too dependent on AI for social interaction, potentially leading to feelings of isolation and anxiety when the AI isn't around.[1]

Using AI for parenting also prompts ethical considerations. AI can assist with education and social tasks, but it can't replace the care and nuance that only human parents or guardians can provide. Businesses adopting AI must balance these benefits with responsibility, considering the ethical and psychological impacts for the sake of genuine human connections and promoting overall well-being.

SEO and Content Optimization

If you're running a business website, blog, or really any online content nowadays, you simply can't afford to ignore AI-driven SEO and content optimization. Keep in mind that these aren't simply the silly AI writing tools that spin spammy clickbait, either. AI can supercharge your entire content strategy from start to finish in incredibly sophisticated ways.

Let's start with the research and ideation phase for new content pieces. Rather than guessing blindly about what topics your audience might find interesting based on gut instinct, you can leverage AI natural language processing models to automatically analyze everything from search query data to social media chatter, forums, reviews, competitor intel, and more. The AI can rapidly pinpoint emerging trends, content gaps, and underserved content niches your editorial calendar should be pursuing.

For example, let's say you're a travel brand and want to create fresh destination content to drive bookings. By ingesting gigabytes of publicly available traveler reviews, online search data, and social media commentary, an AI system could identify that sustainable eco-tourism and off-the-beaten-path experiences have been surging in interest lately. It could then cross-reference that customer demand against your existing inventory of travel packages and content to spot prime opportunities for creating optimized landing pages, long-form guides, and other SEO content targeting those hot keyword topics.

And speaking of keywords, many may not realize that we've come a long way from the brute force, robotic "keyword stuffing" tactics of old-school SEO. With natural language AI models now advanced enough to grasp contextual semantics and human communication, the entire game has shifted to delighting humans first while secondarily signaling relevance to search engines. That means structuring content to flow naturally while incorporating relevant keywords and related terms organically based on a semantic understanding of the core topic.

AI writing tools can even automatically generate first drafts, complete with keyword mentions working in context. The human writers and editors then kick in to refine those AI-generated outputs into clear, authoritative content that ticks both the human readability and SEO boxes. With AI handling a lot of the heavy keyword and topic modeling lifting up front, human writers get to focus on higher-level creative priorities.

But of course, there's far more to the SEO content optimization story than just keywords. With multimodal AI systems now able to analyze the full spectrum of text, imagery, audio, and video artifacts associated with any given content experience, the optimization opportunities are virtually limitless.

For text, AI can focus on details like content structure, formatting, entity extractions, reading comprehension levels, and more to determine areas for improving engagement. For visual assets, computer vision models can assess image composition, sharpness, color palettes, emotional resonance, and even specific elements to include or exclude based on what tends to command audience attention most effectively. Similar video and audio intelligence open new frontiers in polishing production value at scale.

Speaking of scale, that's perhaps where AI's automation strengths shine brightest in the world of SEO. Traditionally, content optimization efforts were highly manual lift, with SEO specialists painstakingly auditing rankings, analytics data, page metadata, backlink portfolios, technical factors, and more to detect improvement areas on a site-by-site, page-by-page basis. Now, those AI systems can automate the heavy lifting end to end.

First, they continuously scan sites to build interactive maps of all indexed content, information architecture, metadata, social embeds, and user-generated discussions. The AI can then analyze all those structured and unstructured data graphs to produce granular optimization recommendations. For large enterprises with vast numbers of digital content properties, this AI-powered auditing is an indispensable force multiplier compared to manual efforts.

Once suggested optimizations are identified, many leading SEO platforms will integrate AI-powered workflows to streamline implementation at scale. Rather than webmasters individually remediating meta tags, internal link structures, media compressions, and a million other little details one by one, the AI handles bulk updates through APIs. In fact, they can even run machine learning predictive simulations estimating the potential SEO impact of various changes before deployment.

Behind all these content optimization capabilities sits large language models and multimodal understanding systems trained on just staggering volumes of internet data. Think petabytes of crawled websites, forums, publisher archives, user-generated repositories, video transcripts, and pretty much the entire open web corpus an AI can gobble up and learn from.

With foundations built on such comprehensive knowledge graphs encompassing how content and information spread across the internet, these SEO AI systems can start to intuit the subtle signals that seem to differentiate authoritative, engaging, trustworthy content that ranks well versus the mediocre stuff that drowns on page twenty of the Search Engine Result Pages (SERP). And they aren't just looking for on-page signals either. The AI incorporates reputation factors, entity associations, user behavior cohorts, and even predictive modeling around evergreen vs. ephemeral content lifespans.

To top it off, platforms are increasingly folding in generative AI capabilities to streamline optimized content production right alongside the auditing insights. In one unified system, companies can receive automated SEO recommendations, draft initial AI-powered content, fine-tune it based on readability and media enrichment suggestions, and then streamline large-scale deployment based on predicted business impact estimates.

Chatbots and Virtual Assistance for Customer service and Engagement

Rather than having to click through endless phone trees and web forms to find basic information, customers can simply chat or speak their requests to an always-available virtual agent. AI assistants use advanced natural language processing models to accurately interpret colloquial phrasing, extract key intents, and quickly provide relevant solutions—all while maintaining an empathetic, on-brand voice for that particular business. For example, queries as varied as "How do I reset my password?" or "My widget seems to be malfunctioning after the latest update." They can even be completely open-ended and up to interpretation. For example, consider the rather unspecific query: "I'm having issues setting up my new product." These can all be accurately handled without frustrating customers or shuttling them through redirect hellscapes. The AI handles triaging issue areas, offers troubleshooting steps or self-service flows based on deep integration into the company's knowledge bases, and can seamlessly escalate to live support if needed. All within a conversational, natural interaction loop.

Rigid curated exchanges or robotic responses devoid of any emotional intelligence are over. The best virtual assistants can pick up on conversational subtext, sentiment signals, and personal rapport and even adapt their own personalities to what resonates best with that individual customer's traits. So the same bot could shape-shift between a warm, empathetic voice for a frustrated user, embody casual and playful for younger demographics, or clinical and just-the-facts for an old-school, no-nonsense persona.

What's interesting is just how advanced the underlying AI language models have become at developing that level of context-aware communication. The latest large language models have essentially ingested the collective knowledge of how humans communicate across websites, books, videos, and more to understand semantic meaning, follow complex conversation flow, and generate intelligent, relevant responses on the fly.

Add in their ability to integrate real-time with business data sources like order systems, billing platforms, user profiles, and CRMs, and suddenly, you have virtual agents that can answer hyper-personalized inquiries tailored to each customer's unique circumstances and history. That same model could advise on everything from return policies and account lookups to curating dynamic product bundles and purchase recommendations based on their datasets.

Even better, many of these AI assistants are also multimodal, able to ingest visual inputs like screenshots or photos customers share to get a richer context. Rather than struggling to describe a buggy experience over chat, you could just snap a photo, which the computer vision model analyzes to isolate the issue before recommending solutions.

And that's just the start. Companies are increasingly leveraging conversational AI platforms as multipurpose digital ambassadors across all their customer touchpoints beyond just troubleshooting and support. They act as sales assistants walking prospects through the funnel, friendly onboarding guides for new customers, and even proactive engagements surfacing personalized advice, offers, or retention nudges based on predictive churn models.

For financial service firms, virtual agents handle complex queries around banking activity, loans, investment portfolios, and tax scenarios using their in-depth data-crunching capabilities. Some retail trailblazers are even experimenting with 3D-embodied chatbots that merge physical and virtual worlds into immersive shopping experiences.

While consumer use cases tend to get most of the buzz, conversational AI is also rapidly being deployed in business-to-business settings, particularly contact centers and customer support operations. In addition to handling that first line of customer requests and triaging, these systems are augmenting human agents too in real-time by surfacing recommended responses based on dialog context, pulling in relevant information from knowledge bases, and even analyzing tone and sentiment to suggest de-escalation strategies. They're enabling a best-of-both-worlds experience where customers still get that warm human touch for sensitive issues and complex scenarios but with much of the busy work offloaded to virtual assistants behind the scenes.

Of course, deploying robust conversational AI at an enterprise scale is no trivial undertaking. Businesses need to carefully design speech and text pipelines to process all those digital interactions, map out intricate dialog workflows across use cases, and compile comprehensive training datasets covering their specific product catalogs, process knowledgebases, and customer scenarios. Ongoing monitoring is critical, too, iterating models over time as language evolves, business offerings change, and edge cases present themselves.

Voice models need specialized training corpora to handle things like multilingual support, accents, background noise, and audio fidelity fluctuations. Security and privacy controls must be watertight. Short of proper governance, these AI systems can quickly go haywire with toxic responses and biases if not responsibly developed with safeguards around factual integrity, ethical boundaries, and inclusive training practices.

Get it right, and the rewards of mastering conversational AI could pay massive dividends in deepening customer relationships. In that sense, they become brand-builders and loyalty drivers rather than just cost centers.

Predictive Analytics in Sales Forecasting

Onto sales forecasting. Let's say you're running a bustling e-commerce business. You have thousands of products, a global customer base, and more sales data than you know what to do with. Traditionally, you might rely on your sales team's instincts or basic historical trends to guess future demand. That doesn't cut it anymore.

AI ingests all historical sales data, current market trends, economic indicators, weather patterns, social media sentiment, and a whole host of other variables to build incredibly nuanced forecasts. In practice, say you're a clothing retailer gearing up for autumn. Your predictive analytics system might first analyze past sales data to identify seasonal trends. But it will also factor in:

- Weather forecasts (A warmer-than-average autumn might delay sweater sales)
- Economic indicators (Consumer confidence levels can impact discretionary spending)
- Social media buzz (Is there sudden hype around a particular style or influencer collab?)
- Competitor activity (Are they running major promotions that could impact your sales?)
- Supply chain data (Are there potential disruptions that could affect product availability?)

This gets fed into complex machine-learning algorithms that spot patterns and correlations humans might miss. The resulting forecasts are highly specific, offering insights such as "a projected 15%increase in sales of men's corduroy jackets in the Northeast region during the third week of October, driven primarily by cooler temperatures and a trending TikTok fashion challenge."

These systems can also optimize your whole sales funnel. For instance, they might identify that customers who purchase a certain combination of items are 30 percent more likely to become repeat buyers. With that insight, you could create bundle offers or personalized follow-up campaigns to boost customer lifetime value.

Predictive analytics dynamically adjust inventory predictions using real-time data, so you avoid both costly stockouts and excess inventory. Let's say you're an electronics retailer. Your predictive model might notice a surge in social media chatter about a particular gaming console that's set to release a highly anticipated new title. It could automatically flag this trend, cross-reference it with your current stock levels and supplier lead times, and recommend increasing your order quantity to meet the expected demand spike. However, these systems are constantly learning and refining their models based on new data. So, if that gaming console doesn't sell quite as well as expected, the system will analyze why its prediction was off and use that information to improve future forecasts.

Now, you might be thinking, "This all sounds great for big companies with massive datasets, but what about smaller businesses?" Luckily, predictive analytics are increasingly accessible to businesses of all sizes. Imagine you run a neighborhood coffee shop. A basic predictive analytics tool could help you forecast daily foot traffic based on factors like weather, local events, and historical patterns to optimize staffing levels and reduce waste from over-preparing perishable items.

Implementing predictive analytics effectively does come with challenges. Data quality is at the center. You need to input data that's clean, consistent, and comprehensive. That's why the most successful implementations of predictive analytics in sales forecasting tend to be hybrid approaches. AI does the heavy lifting of crunching numbers and spotting patterns, but human experts are still involved in interpreting results, providing context, and making final decisions. Your predictive model might flag an unexpected dip in forecasted sales for a usually reliable product line. A seasoned sales manager might recognize that this coincides with a competitor's product launch and can adjust the strategy accordingly.

Lead Generation and Qualification

In simple terms, lead generation and qualification mean finding potential customers (leads) who might be interested in your product or service and then figuring out which ones are worth pursuing (qualification). It used to be a pretty hit-or-miss affair, but AI is changing it in some mind-blowing ways.

Traditionally, lead generation might involve setting up a booth at a trade show, running generic online ads, or good old-fashioned networking. However, AI can now scour the internet for signals that someone might be in the market for what you're selling: social media posts, website visits, job applications, and even company financial reports. For example, you're selling high-end cybersecurity to enterprise clients. An AI system might flag a company as a potential lead if it detects:

- Recent news articles about the company expanding into new markets (they might need upgraded security)
- LinkedIn updates showing they've hired a new CTO (fresh decision-maker who might be open to new solutions)
- Increased visits to cybersecurity-related websites from their corporate IP addresses (indicating growing interest in the topic)
- A spike in social media mentions about data privacy concerns (suggesting they're thinking about security issues)

All of this happens automatically, creating a constant stream of potential leads that are way more trained than anything you'd get from traditional methods.

But generating leads is only half the battle. The real magic happens in the qualification process. Instead of relying on gut feelings or simple scoring systems, AI can analyze hundreds of data points to predict which leads are most likely to convert into customers.

To break down a real-world example, say you're a B2B software company selling a project management tool. Your AI lead qualification system might consider:

- Company size and industry (gleaned from public databases)
- Recent funding rounds or financial performance (indicating available budget)
- Current tech stack (are they using complementary or competing tools?)
- Engagement with your marketing content (website visits, email opens, webinar attendance)
- Social media sentiment toward project management topics
- Hiring patterns (are they growing their project management team?)

AI crunches all this data and assigns each lead a probability score for conversion. It also recommends the best approach for each lead. Maybe one company responds well to technical deep dives, while another prefers high-level ROI discussions. AI can suggest talking points, content to share, and the best time to reach out.

Better yet, many CRM platforms now offer built-in AI features that can work with relatively small datasets. A local real estate agency might use a basic AI tool to analyze its client database and identify patterns in who tends to buy or sell homes. The system might discover that people who have recently changed jobs or had a child are more likely to be in the market for a new house.

A chatbot on a B2B software website might first ask about the visitor's role and company size. Based on the responses, it could then pose more specific questions about pain points and current solutions. The AI analyzes the responses in real time and adjusts its line of questioning to gather the most relevant information. By the time a human sales rep gets involved, they have a wealth of qualification data to work with.

There's also a growing trend toward using AI for "account-based marketing" (ABM) in the B2B space. Instead of casting a wide net, ABM focuses on identifying and targeting specific high-value accounts. AI helps by analyzing vast amounts of data to identify companies that fit your ideal customer profile, even if they haven't explicitly shown interest yet.

Personalized Marketing at Scale

Personalized marketing at scale is about using data and AI to create unique, relevant experiences for each customer across all touchpoints. Take Amazon, for example. Their recommendation engine does not just show you products similar to what you've bought before, but AI also analyzes your browsing history, purchase patterns, wish list items, the time you spend looking at certain products, and more. It then cross-references this with data from millions of other users to predict what you're likely to buy next. Some estimates suggest that up to 35 percent of Amazon's revenue comes from these recommendations.

Major corporations aside for a moment, you may wish to know how a smaller business might use these techniques. With AI-powered personalization, someone running a local gym could:

- Customize a website for each visitor. A 20-something who visited your CrossFit page might see images of high-intensity workouts, while a retiree might see content about low-impact classes and senior fitness programs.

- Instead of blasting everyone with the same "" email in January, you could tailor messages based on each member's attendance patterns, favorite classes, and fitness goals.

- Create dynamic pricing models. Maybe you offer a discount on personal training sessions to members who haven't visited in a while, or a special rate on nutrition counseling to those who've been consistently hitting the gym.

- Optimize a mobile app. If a user always checks in for the 6 AM yoga class, their app home screen could show a quick "Check In" button for that class right at the top.

On a larger scale, consider Netflix personalizes almost every aspect of the user experience. The artwork you see for each show, the order of the rows on your homepage, and even the trailers you see are often edited differently based on your viewing history.

For example, if you watch a lot of romantic comedies, the thumbnail for *Stranger Things* might show a moment between two characters rather than a scary monster. If you're into action movies, you might see a more intense, thriller-style trailer for the same show. All of this is controlled by AI systems that are constantly testing and learning what works best for each user.

AI may also determine the best time to send an email, the optimal frequency for push notifications, and, in some cases, the tone of voice that resonates best with each customer. Spotify uses AI to create playlists that reflect just this, but they also use it to determine when to introduce new music to listeners. If you always listen to upbeat pop on your morning commute, that's when they'll slip in new, similar tracks you might like.

When implementing this in your own business, take a platform like Optimizely. It allows you to run A/B tests on your website, but with AI, it goes way beyond simple "version A vs version B" testing. You can create dozens of variations of a page, and the AI will automatically serve the best version to each visitor based on what it knows about them.

Or consider a tool like Persado, which uses AI to generate and optimize marketing copy. It can create thousands of variations of an email subject line or ad headline, test them in real time, and learn which types of language resonate best with different audience segments.

Stitch Fix, the online personal styling service, uses AI to predict not just what clothes a customer might like now, but also what they might like in the future. Their algorithms analyze purchase history, broader fashion trends, seasonal changes, and how a customer's style evolves.

Enhancing CRM with AI

Traditional CRM systems are great for storing customer data, tracking interactions, and managing pipelines. But add AI to the mix, and suddenly, your CRM becomes a proactive powerhouse that can spot opportunities, flag potential issues, and guide your team toward the most effective actions.

Take Salesforce's Einstein AI, for example. One of its coolest tricks is lead scoring. Instead of relying on simple rules like "assign ten points if they downloaded a whitepaper," Einstein analyzes hundreds of factors to predict which leads are most likely to convert. It looks at things like email engagement, website behavior, social media activity, and news about the company to build a holistic picture of each lead's potential.

Einstein can also recommend the next best action for each lead. Maybe it's noticed that leads from a certain industry tend to convert better after attending a webinar. Or perhaps it shows that sending a specific piece of content at just the right time in the sales cycle boosts close rates.

Now, let's talk about customer service. AI-powered CRMs like Zendesk's Answer Bot use natural language processing to understand customer questions and can automatically provide relevant answers from your knowledge base, and it learns from every interaction. If it notices that certain answers aren't resolving issues, it will flag those for human review. Over time, it gets better and better at handling complex queries, freeing up your human agents to tackle the trickier issues.

So, what about those human agents? AI is making the agents more effective. A system that can listen in on customer calls in real-time, analyzing both what's being said and how it's being said can then prompt the agent with relevant information, suggest effective responses, or flag if the customer's tone indicates they're getting frustrated. Companies like Cogito are already deploying this technology, helping agents navigate difficult conversations with empathy and precision.

Let's shift gears and talk about predictive analytics. This is where AI-enhanced CRMs start to feel almost like crystal balls. They analyze historical data to forecast everything from sales projections to customer churn risk. The CRM might notice that customers who haven't logged into your software for thirty days and haven't opened your last three emails are at high risk of canceling their subscription. It can then automatically trigger a re-engagement campaign or alert an account manager to reach out. If you're a B2B software company, your AI CRM might notice that companies in a certain industry tend to dramatically increase their usage of your product about three months before they go through a funding round. Knowing this, your sales team can proactively reach out to offer expanded services or support.

However, keep in mind that if your CRM is full of outdated or inaccurate information, even the smartest AI will struggle to generate useful insights. That's why many companies are investing in data cleaning and enrichment services as a first step toward AI implementation. There's also the question of user adoption. Sales reps and customer service agents who are familiar with traditional CRM systems might be skeptical of AI-powered features. It's best to focus on user experience and demonstrable value. When a sales rep sees that the AI-recommended email template consistently gets higher engagement rates or that the 'high potential' leads are closing at a notably higher rate, any initial skepticism quickly diminishes.

Case Studies

Netflix is a standout example of AI's power in marketing. It employs sophisticated machine learning algorithms to analyze user data, such as viewing habits, search history, and the time-of-day content consumed. For example, if a user binge-watches a certain genre, Netflix's AI will suggest similar shows or movies that align with their preferences[2].

Sephora, the cosmetics retailer, has embraced AI to enhance customer experience. Sephora's Virtual Artist uses AI and facial recognition to allow customers to try different makeup products virtually. Customers can upload a photo and see how various products look on their faces. This tool has significantly boosted customer engagement and satisfaction, leading to higher sales.

Coca-Cola uses AI to quickly create personalized marketing content based on consumer data like buying habits and social media activity. This has boosted engagement by delivering content that fits audience preferences. AI also cuts production time and costs, freeing up resources for other marketing initiatives.[3]

Starbucks employs AI through the Deep Brew AI engine, which analyzes data from its app and loyalty program to personalize customer interactions. Deep Brew uses machine learning to understand customer preferences and habits. This personalization has led to increased sales, higher average transaction values, and improved customer loyalty. If a customer frequently orders a particular type of coffee in the morning, Starbucks might send a promotion for a complimentary pastry to go with it.

Volkswagen is using the power of AI to enhance marketing. By using AI to automate its ad-buying decisions, Volkswagen has cut costs and increased dealership sales by 20 percent. The AI system analyzes social media data to create personalized and engaging content for its audience.

Euroflorist, a leading European online florist, adopted an AI-driven approach to website optimization, using the power of massively multivariate testing. This involved using AI platforms to test thousands of website variations, analyze user interactions, and make data-driven decisions about design and content. The result was a 4.3 percent increase in conversion rates and a more user-friendly shopping experience.

---

1. Chaturvedi, R., Verma, S., Das, R., & Dwivedi, Y. K. (2023). Social companionship with artificial intelligence: Recent trends and future avenues. *Technological Forecasting and Social Change, 193*, 122634. https://doi.org/10.1016/j.techfore.2023.122634

2. Huang, M. H., & Rust, R. T. (2020). A strategic framework for artificial intelligence in marketing. *Journal of the Academy of Marketing Science, 49*(1), 30-50. https://doi.org/10.1007/s11747-020-00749-9

3. DigitalDefynd. (2024). 10 Successful AI Marketing Campaigns & Case Studies. DigitalDefynd. Retrieved from https://digitaldefynd.com/IQ/ai-marketing-campaigns/

Chapter Seven

AI in Operations, Logistics and Supply Chain Management

This chapter will address how Artificial Intelligence affects how companies move products from concept to customer. No more static supply chains and gut-feeling inventory management. AI is creating an era of unprecedented efficiency, prediction, and automation. We'll explore how AI is transforming every aspect of operations—from tracking goods across the globe to optimizing warehouse layouts, from predicting demand fluctuations to automating procurement.

Tracking and Monitoring Supply Chain Activities

Imagine you're running a global electronics company. Your products are assembled in China, using components from a dozen different countries, and then shipped to distribution centers around the world. Not so much with AI.

AI now analyzes vast amounts of data to predict supply and demand fluctuations. IBM's Watson Supply Chain uses natural language processing to scan news reports, weather forecasts, and social media chatter. It might notice that a drought in Taiwan could lead to a shortage of semiconductor chips six months down the line. A global electronics company can line up alternative suppliers or adjust production schedules before competitors even know there's a problem.

Furthermore, AI can analyze your purchasing patterns alongside market data to suggest optimal buying times. Maybe it notices that the price of a key component tends to dip every year just after a major industry conference. That's the kind of pattern that might be too subtle for a human to spot but obvious for an AI crunching through years of data'.

Now, let's move on to the manufacturing process. This is where Internet of Things (IoT) devices and AI come into play. If every machine in your factory uses sensors feeding data to a central AI, it can predict when a machine is likely to break down based on subtle changes in its performance metrics.

Siemens has implemented what they call a "digital twin" system in their electronics plants. It's a virtual replica of the entire production line, updated in real-time with data from IoT sensors. AI can run simulations on this digital twin to optimize production flow, predict maintenance needs, and test out new configurations without disrupting actual production.

The real magic happens when you start connecting these systems across your entire supply chain. Suddenly, a delay at your component supplier can automatically trigger adjustments in your production schedule and shipping plans.

Traditional route planning was based on static factors like distance and average traffic conditions. However, AI systems also factor in real-time data on weather, traffic, port congestion, and even social events that might cause delays. In practice, DHL has implemented an AI system called IDEA (Intelligent Delivery Execution Assistant) that can predict delivery delays up to ten days in advance. It analyzes over fifty different parameters, including historical data, current shipment information, and external factors like weather and traffic.

AI can also optimize routes in ways humans never could. Consider the classic "traveling salesman problem" of finding the most efficient route between multiple points. When dealing with hundreds of delivery stops across a city, the number of possible routes becomes astronomically large. AI algorithms can solve these complex optimization problems in seconds, potentially saving millions in fuel costs and reducing delivery times.

Predictive analytics for Forecasting and Inventory Management

Picture yourself as a fashion retailer trying to predict demand for swimwear. A traditional system might simply look at last year's sales and add a percentage for growth. But a sophisticated predictive analytics system will also factor in things like long-range weather forecasts (a cooler summer could mean lower demand), travel trends (are more people planning 'staycations' this year?), and even social media buzz around certain styles or brands.

Let's also look at how these systems handle the tricky world of new product launches. Traditionally, forecasting demand for a new product was more art than science, often relying heavily on market research and educated guesses. But predictive analytics is changing that game too.

For example, when Nintendo launched its Switch console, it used predictive analytics to forecast demand. Their system analyzed pre-order data, social media buzz, and search engine trends to predict overall demand, regional variations, and potential supply constraints. That way, they could adjust their production and distribution strategies in real time and avoid the supply shortages that had plagued previous console launches.

Furthermore, predictive analytics isn't just about forecasting demand. It can also substantially optimize your inventory management. These systems may calculate optimal reorder points and suggest ideal safety stock levels. When necessary, they also predict potential supply chain disruptions. So, if the system detects a potential shortage of a key ingredient, it can automatically suggest alternative suppliers or adjust production schedules to mitigate the impact.

AI in Route - Optimization and Fleet Management

Someone running a delivery company with a fleet of 100 vehicles, each making fifty stops per day, deals with 5,000 deliveries daily. Now, factor in traffic patterns, delivery time windows, vehicle capacities, and driver schedules. Suddenly, you're dealing with an astronomical number of possible route combinations.

Traditional methods relied on predetermined routes or basic algorithms that couldn't account for real-time changes. AI, however, can process vast amounts of data in seconds, considering traffic conditions, weather, possible construction works, customer preferences, and more. For example, UPS implemented an AI-powered tool called ORION (On-Road Integrated Optimization and Navigation) back in 2012. This system analyzes up to 200,000 options for each route.

Moreover, AI systems can prioritize routes based on various factors. Take the example of a grocery delivery service during the COVID-19 pandemic. The AI could prioritize deliveries to elderly or immunocompromised customers, ensuring they received their essentials first.

Now, let's talk about AI in fleet management. AI-powered predictive maintenance systems use sensor data from vehicles to predict when a part is likely to fail, allowing for proactive repairs and reducing costly breakdowns. It can analyze driving behavior, providing insights that improve safety and efficiency.

Moreover, AI is helping companies optimize their fleet composition. By analyzing historical data and future projections, AI can recommend the ideal mix of vehicle types for a company's needs. Maybe you don't need as many large trucks as you thought, or perhaps electric vehicles would be more cost-effective on certain routes. AI can crunch these numbers and provide data-driven recommendations.

AI is also making waves in the realm of autonomous vehicles. While fully self-driving trucks aren't commonplace yet, AI-assisted driving is already here. This technology promises to address the chronic driver shortage in the trucking industry while potentially improving safety and efficiency.

Automated Warehousing Solutions

Picture a massive warehouse, but instead of workers on forklifts, there is an army of robots picking, packing, and moving products. This is happening right now in warehouses across the globe. Amazon has deployed over 350,000 mobile drive unit robots in their fulfillment centers. These nifty little machines are moving entire shelves of products to human workers for picking and packing.

Automated storage and retrieval systems (AS/RS) turn warehouses into high-tech vaults of efficiency. They use cranes that run on rails to store and retrieve products from tall storage racks. These robots can pick up and move crates of groceries, bringing them to human workers for final packing.

But what about the actual picking of individual items? That's where some of the most exciting innovations are happening. Boston Dynamics, the company famous for its slightly unnerving dog-like robots, has developed a robot called Stretch. This robot can unload trucks and move boxes around warehouses. It has a powerful suction cup arm that can grab and move boxes weighing up to 50 pounds.

Meanwhile, companies like RightHand Robotics are tackling the challenge of picking individual items from bins. Their RightPick system uses a combination of cameras, machine learning, and a versatile gripper to identify and pick up a wide variety of items. It can handle everything from soft plush toys to rigid boxes, adapting grip strength and technique for each item.

What's better is that many automated warehousing solutions work alongside humans, enhancing their capabilities. Exoskeletons, like those developed by German Bionic, are a prime example. These suits can help workers lift heavy objects with less strain, reducing fatigue and the risk of injury.

Voice-directed warehousing is another technology that's enhancing human capabilities. Systems like Vocollect by Honeywell use voice recognition and text-to-speech to guide workers through their tasks and keep their hands and eyes free. Workers wear a headset that tells them where to go and what to pick, and they confirm their actions by speaking back to the system.

AI for Smarter Procurement

One of the most impressive applications of AI in procurement is in spend analysis. Traditionally, this was a time-consuming task that involved manually categorizing and analyzing purchase data. Now, AI tools do this in a fraction of the time and with far greater accuracy. AI can understand context and nuance, correctly categorizing items even when they're described in different ways.

In supplier selection and management, traditional methods of supplier evaluation often relied heavily on price comparisons and basic performance metrics. AI, however, can analyze a much broader range of factors. For example, Tealbook, a supplier intelligence platform, uses AI to create comprehensive supplier profiles by aggregating data from millions of sources. Their system can evaluate suppliers based on financial stability, sustainability practices, and social media sentiment.

Furthermore, cognitive procurement advisors, like IBM's Watson Supply Chain Insights, may be used to analyze vast amounts of data to provide negotiation recommendations. It's a way to find opportunities for cost savings and optimal order quantities, while also predicting how suppliers might respond to different negotiation strategies.

AI is also helping companies make more ethical and sustainable procurement decisions. For instance, Sourcemap uses AI to analyze supply chains and identify potential risks, including environmental issues and labor violations.

One of the most exciting developments in AI-powered procurement is the rise of autonomous purchasing systems. These systems can automatically reorder supplies when stock runs low, adjusting order quantities based on predicted demand. Amazon's Dash Replenishment Service takes this concept to the consumer level, allowing smart devices like printers to automatically reorder ink when they're running low. Now imagine this scaled up to an industrial level.

Production Scheduling and Capacity Planning

In the past, production scheduling was usually done using relatively simple methods like Gantt charts or basic software. Unfortunately, these methods might struggle with complex manufacturing. Enter AI scheduling systems that can juggle multiple variables simultaneously, optimizing for efficiency, cost, and quality all at once. An AI system can create detailed production schedules that consider machine capacity, worker availability, material constraints, energy costs, and many other factors.

However, it's not just about creating a perfect schedule and calling it a day. In the real world, things change constantly. A supplier might be late with a delivery, a machine might break down, or you might get a rush order from an important customer. AI systems can quickly adjust schedules in real time, minimizing disruptions and keeping production on track. If there's a roadblock, it will identify a new route instantly.

Now, let's talk about capacity planning, which is about making sure that you have the right resources available to meet demand. It's a tricky balancing act. Too little capacity and you'll miss out on sales, but too much, and you're wasting money on idle resources. Here, AI enables much more accurate demand forecasting. This is where digital twin technology comes in. A digital twin is a virtual replica of your production process that you can use to simulate different scenarios.

General Electric (GE) has been a pioneer in this area. They use digital twins to optimize the production of jet engines. By creating a virtual model of their production line, GE can test different configurations and production schedules without disrupting actual operations.

AI is also helping companies get smarter about how they use their existing capacity. It is useful to analyze data from multiple sources to spot inefficiencies that people might miss. One particularly cool application of AI in production scheduling and capacity planning is in predictive maintenance. BMW uses AI to monitor production lines, cutting down on downtime and improving overall efficiency. Meanwhile, in the energy sector, companies like Enel use AI to keep an eye on wind turbines, saving on maintenance while boosting reliability. By analyzing data from sensors on manufacturing equipment, AI can predict when a machine is likely to break down.[1]

Case Studies

Walmart and Unilever are great examples of how AI is transforming supply chain management. Walmart has integrated AI across various facets of its supply chain, like automation, procurement, and customer interaction. Walmart uses AI for contract negotiations, which results in significant savings and extended payment terms. They've employed AI models such as Google's BERT and GPT-4 for forecasting, procurement, and inventory management. This has improved unit cost averages by around 20 percent. Additionally, Walmart uses AI chatbots to enhance customer care.[2]

Unilever, on the other hand, uses it to drive sustainability and efficiency. Unilever uses AI for inventory management and satellite imaging to enhance traceability and transparency, aiming for a deforestation-free supply chain. Collaborating with Google Cloud and biotechnology specialists, Unilever has introduced energy-efficient vending machines and reformulated products to be more sustainable.[3]

---

1. Boston Consulting Group. (2023). Predictive maintenance in manufacturing: Enhancing efficiency with AI. Retrieved from https://www.bcg.com/publications/2023/predicitive-maintenance-in-manufacturing

2. AIX. (2023). Case Study: Walmart's AI-Enhanced Supply Chain Operations. AI Expert Network. Retrieved from https://aiexpert.network/case-study-walmarts-ai-enhanced-supply-chain-operations

3. AIX. (2023). Case Study: Unilever's Integration of AI in the Supply Chain. AI Expert Network. Retrieved from https://aiexpert.network/case-study-unilevers-integration-of-ai-in-the-supply-chain

Chapter Eight

AI in Human Resources

The following chapter will show you how AI is revolutionizing recruitment, making it faster, fairer, and more effective. We'll explore how it's personalizing onboarding and training, giving each employee exactly what they need to succeed. You will also get a chance to examine the ways AI impacts performance management through continuous feedback and insights. And we'll look at how it's streamlining HR administration, freeing up HR professionals to focus on strategic initiatives.

Recruitment and Talent Acquisition

What better place to start than with the very beginning of the recruitment process: job posting? AI is now being used to write more effective and more inclusive job descriptions. Tools like Textio use natural language processing to analyze job postings and suggest improvements. For example, it might flag words that tend to deter certain groups of applicants and suggest alternatives.

Writing a great job ad is just the first step. Can you imagine sorting through a million job applications a year manually? AI could make it easier by analyzing video interviews. It looks at facial expressions, word choice, mannerisms, and tone of voice to assess candidates. Of course, video analysis is just one way AI is being used to screen candidates. Many companies are using AI chatbots to conduct initial interviews. They work like friendly, tireless recruiters available 24/7 to chat with potential hires.

Many candidates have mixed feelings about AI in hiring, and many don't even realize AI might be involved. While AI can make the process quicker, it also raises concerns about fairness and transparency. For example, research shows that around 66% of U.S. adults feel uneasy about AI making hiring decisions. People worry that AI might misinterpret things like facial expressions or tone of voice, especially when cultural differences come into play.[1]

On the bright side, AI can also find candidates you might otherwise miss. Talent scout tools like LinkedIn Recruiter use AI to help companies find passive candidates—people who aren't actively job hunting but might be perfect for a role. AI analyzes profiles and activity patterns to find people open to new opportunities.

Now, let's talk about biased hiring. It's a huge issue and one where AI could make a real difference or make things worse. Things like a candidate's name, school, or past employers can trigger unconscious biases, leading to unfair treatment. To make AI work for good and avoid perpetuating biases, companies use AI systems that implement neuroscience games to evaluate candidates' cognitive and emotional traits rather than relying on resumes, which can often be skewed. This approach focuses more on the candidate's actual skills and behaviors beneficial for the role rather than details that might not even matter. But it's still necessary to keep checking these AI systems to make sure they're actually reducing bias, not letting them sneak through the system's defenses.

Employee Onboarding and Training

Remember your first day at a new job and all the paperwork, the introductions, the information overload? AI is changing all that. It can answer questions about company policies, help with task prioritization, suggest people to connect with based on the employee's role and interests, and more. The AI tool becomes a friendly AI colleague who knows everything about the company and is always happy to help.

Artificial intelligence is also being used to personalize onboarding. Enboarder, an onboarding platform, uses AI to create customized onboarding journeys for each new hire. The AI considers factors like the employee's role, location, and learning style to make each onboarding unique.

Old-school training methods, more often than not, follow a one-size-fits-all approach. However, an AI system uses machine learning to analyze an employee's performance, knowledge gaps, and learning preferences, then serves up a bite-sized training module. It knows exactly what you need to learn and how you learn best.

Virtual reality (VR) and augmented reality (AR) are also getting in on the action, with AI making these technologies even more powerful for training. Walmart uses VR for training store employees. The system, developed with Strivr, analyzes how employees perform in virtual scenarios and provides personalized feedback.

But AI isn't just about delivering training content; it's also about measuring its effectiveness. Tools like Watershed use AI to analyze training data and provide insights into how well employees are learning and applying their new skills. It can even predict which employees might need additional support before they start to struggle.

Language learning is another area where AI is making a big impact. Duolingo, while not specifically for employee training, shows how AI can revolutionize language learning. The AI adapts to each learner's pace and style. Now, imagine applying this to corporate language training programs. It's like having a polyglot tutor available 24/7 for every employee who needs to brush up on their business English or learn Mandarin for an overseas assignment.

Now, let's be honest: compliance training often struggles to keep people's attention. But AI can create adaptive compliance training that feels more like a game than a chore by adjusting the difficulty and focus of the training based on each employee's responses, without wasting time on what they already know.

Of course, with AI systems collecting and analyzing so much data about employees, companies need to be very careful about how they handle this information. There's also the challenge of employee acceptance. Some workers might feel uncomfortable with the idea of an AI system tracking their learning progress or giving them feedback. As a response, companies must clearly explain what they track and why to avoid making workers feel overly monitored. They could also invite employees to share feedback on how these systems affect their work, helping address privacy concerns. Moreover, businesses should run regular audits to keep AI tools in check and ensure they're used responsibly within ethical boundaries.

Performance Management

The days of annual performance reviews were a time of the dreaded meetings where a year's worth of work was condensed into a single conversation. However, with AI, the system provides continuous feedback, using natural language processing to analyze communications and project management tools to assess performance in real-time. It can also analyze data from various sources—past performance reviews, project outcomes, and even email sentiment—to predict which employees are likely to succeed in different roles.

AI is changing goal setting too. OKR (Objectives and Key Results) platforms like Gtmhub use AI to help employees set better goals and track their progress. The AI can suggest relevant objectives based on company strategy and an employee's role and predict the likelihood of achieving those goals.

What about bias in performance evaluations? This is a major issue and one where AI is making a surprising difference. Tools like Textio may analyze performance reviews for biased language. They might flag if managers consistently use different language to describe the performance of male and female employees, or those of different ethnicities. Meanwhile, platforms like Degreed use AI to analyze an employee's skills, performance, and career goals to suggest relevant learning opportunities. It might recommend a specific online course, a mentoring relationship, or a stretch assignment without relying on irrelevant traits.

Now, onto teamwork. 'Microsoft has been doing work in this area with the Workplace Analytics tool. It uses machine learning to analyze collaboration patterns—email traffic, meeting schedules, and even the text of digital communications—to identify what makes high-performing teams tick. It can spot things like whether a team is collaborating effectively or if certain members are at risk of burnout.

In fields like sales, AI may provide real-time coaching. Gong.io, for example, uses natural language processing to analyze sales calls. It can identify successful patterns in how top performers communicate with clients and provide suggestions to other team members in real time.

HR Administration

The bread and butter of HR admin is managing employee data. Machine learning algorithms can predict which employees are likely to leave the company, allowing HR to take proactive steps to retain top talent. AI isn't just about predicting the future; it's also about making the present run more smoothly. Chatbots are becoming increasingly common in HR departments, where they handle routine queries and tasks. IBM's Watson has been put to work in HR departments around the world. It can answer employee questions about everything from vacation policies to health benefits 24/7.

Don't forget about one of the most arduous tasks many HR professionals face: scheduling. AI is revolutionizing this, too. Tools like Calendly use AI to automatically find the best times for meetings, interviews, and other appointments, considering everyone's schedules and preferences.

Payroll is another area where AI is making waves. ADP, one of the biggest names in payroll processing, uses AI to detect and prevent payroll errors. The system spots unusual patterns or discrepancies that might indicate a mistake or even fraudulent activity. This is an accountant who never misses a detail and never needs a coffee break.

But what about the more complex aspects of HR admin, like compliance? AI tools can analyze company policies and practices to help them stay compliant with ever-changing labor laws. The ComplianceHR tool can even generate custom policies.

Case Studies

The tech giant IBM is a great example of a company using AI chatbots in employee onboarding. These chatbots answer questions, guide new hires through company policies, and facilitate the completion of necessary paperwork. Natural language processing allows these chatbots to interact with new employees in a human-like, individualistic manner, significantly reducing the time HR professionals spend on these tasks and ensuring new hires receive consistent and accurate information. This approach has resulted in more efficient onboarding and improved employee satisfaction[2].

---

1. Pew Research Center. (2023, April 20). AI in hiring and evaluating workers: What Americans think. *Pew Research Center: Internet & Technology*. Retrieved from https://www.pewresearch.org/internet/2023/04/20/ai-in-hiring-and-evaluating-workers-what-americans-think/

2. IBM Corporation. (n.d.). The business case for AI in HR. Retrieved from https://www.ibm.com/downloads/cas/A5YLEPBR

Chapter Nine

AI for Financial Management and Fraud Detection

Envision a world where financial decisions are made in milliseconds, fraud is caught before it happens, and your accountant is an AI. We're already there. This chapter explores the exciting ways AI is shaking up the finance world, from turbocharging accounting processes to predicting market trends with scary accuracy. We'll explore how it's making payroll a breeze, helping businesses see into the financial future, and outsmarting crafty fraudsters.

AI in Accounting and Financial Reporting

Data entry used to be where interns or junior accountants would spend hours manually inputting figures from invoices and receipts. Now, smart optical character recognition (OCR) systems can scan documents and extract relevant information in seconds, achieving accuracy rates of up to 98 percent. That's not just saving time, it's dramatically reducing human error.

Moreover, auditing methods often rely on random sampling, which can miss subtle patterns of fraud. Enter AI-powered continuous auditing that can analyze 100 percent of transactions in real-time, flagging anomalies that human auditors might miss. For instance, the software may use machine learning algorithms to detect unusual patterns in financial data.

'AI is now venturing into the realm of financial reporting itself. Natural Language Generation (NLG) systems automatically draft financial reports based on raw data. Narrative Science's Quill platform, for example, can generate human-readable narratives from complex financial datasets. It's not just regurgitating numbers but' providing context and insights in the needed language.

Of course, we can't talk about AI in accounting without mentioning the big players. The Big Four accounting firms are all investing heavily in AI. PwC has developed an AI-powered tool called GL.ai that can analyze billions of data points to identify high-risk transactions. Deloitte's Argus system uses natural language processing to 'read' thousands of pages of contracts and extract key information in minutes.

AI in Payroll Administration

AI-powered systems like Replicon and Kronos are making a difference in how businesses monitor employee hours. They use machine learning algorithms to analyze patterns in employee clock-ins and clock-outs, flagging anomalies that might point to either slacking off or honest mistakes. For example, Replicon's AI can detect if an employee consistently clocks in a few minutes late but clocks out exactly on time. Luckily, rather than just being a watchdog', AI is also making life easier for employees. Take chatbots, for example. Companies have integrated AI-powered virtual assistants into payroll systems that can answer common payroll questions, help employees access pay stubs, or assist with updating personal information.

Now, let's get into the nitty-gritty of payroll calculations. Payroll software like ADP's Wisely Pay uses machine learning to automatically calculate complex pay structures, including overtime, shift differentials, and bonuses. It can even factor in local tax laws and regulations, which is highly beneficial for companies operating across multiple states or countries. Payroll systems also use AI to forecast future labor costs using historical data and current trends. For example, a retail chain might use these predictions to optimize staffing levels during peak shopping seasons.

AI is also a champion at catching mistakes before they become problems. UKG Pro's payroll software uses machine learning to analyze historical payroll data and flag potential errors. If an employee's paycheck suddenly includes an unusually large amount of overtime, for example, the system might flag it for review.

Believe it or not, AI is making payroll more human. Paradoxically, by automating routine tasks, AI is freeing up payroll professionals to focus on more strategic, people-centered work. They can spend more time analyzing payroll data to identify trends in employee compensation and satisfaction or take on other humane tasks.

AI in Financial Forecasting and Analytics

Financial forecasting often relies on historical data and basic trend analysis, but AI takes it to a whole new level. Take Salesforce's Einstein Analytics, for example. This smart system analyzes a ton of variables like social media trends, economic indicators, and weather patterns to predict future sales. Furthermore, it understands the present in incredible detail. Machine learning algorithms are sifting through vast amounts of financial data to uncover insights that humans might miss to create more comprehensive future scenarios than humans might.

While AI can't predict the stock market with 100 percent accuracy (if it could, everyone would be a billionaire), it's getting eerily good at spotting trends. Firms use sophisticated AI models to analyze market data, news, and even satellite imagery to inform their trading strategies. These AI-driven quantitative investment funds, also known as "quant" funds, often outperform human-managed ones by a significant margin[1].

Thanks to more accessible AI tools, small businesses are getting in on the action, too. Fathom, for example, is an AI-powered financial analysis tool that helps small businesses make sense of their financial data. It can automatically generate financial reports, highlight key performance indicators, and suggest areas for improvement.

'Apps like Mint and Personal Capital use machine learning to analyze your spending habits, predict future expenses, and offer personalized financial advice. These robo-advisors are making sophisticated analysis available to everyone, not just those who can afford a human financial advisor.

AI is starting to factor in non-traditional data sources for financial forecasting. For example, some hedge funds are using natural language processing to analyze social media sentiment about companies. Imagine an algorithm that can read millions of tweets and predict how people feel about a brand. That's powerful for forecasting stock prices or consumer behavior.

Applications in Risk Management

Credit risk assessment is where AI shows off its power fantastically. It no longer needs to rely heavily on a handful of factors like credit history and income but instead analyzes hundreds of data points to get a more accurate picture of creditworthiness. Upstart's AI model considers over 1,000 variables and even non-traditional data like education and job history to approve more borrowers than traditional models while keeping loss rates the same.

However, credit risk is just the tip of the iceberg. Imagine running a massive factory with thousands of moving parts. How do you predict when a piece of equipment might fail? Companies are using predictive maintenance to analyze sensor data from industrial equipment and detect subtle changes in vibration or sound that might indicate a future failure, often weeks or months in advance. That's preventing potential safety hazards while also saving money.

Now, it's worth shifting attention to a risk that keeps business leaders up at night: cybersecurity. With cyber-attacks becoming increasingly sophisticated, rule-based security systems just can't keep up. However, machine learning created self-learning security systems to establish a "pattern of life" for every user and device in a network, then detect and respond to anomalies instantly. It may detect and stop a cyber-attack at a major bank in seconds before any data is compromised.

Let's not forget about market risk. In the world of trading, seconds can mean the difference between profit and loss. That's why firms like Two Sigma use AI to analyze market data in real-time. Algorithms can process news, social media sentiment, and market movements to make split-second trading decisions.

Even supply chain risk management is getting an AI makeover. Companies like LLamasoft use machine learning to simulate thousands of potential supply chain disruptions—from natural disasters to geopolitical events—and recommend optimal strategies. During the COVID-19 pandemic, AI helped companies quickly adjust their supply chains to cope with unprecedented disruptions.

Fraud Detection and Prevention

Credit card fraud is a massive (and ever-growing) problem. So, AI is fighting back and winning. Mastercard's Decision Intelligence technology uses machine learning to analyze over 200 data points for every single transaction, such as spending habits or location data. The result is reduced false declines while still catching the bad actors. It's so fast that it can approve or decline a transaction in milliseconds.

The old method of manually reviewing claims to detect insurance fraud is like trying to find a needle in a haystack. Machine learning analyzes thousands of claims in seconds, flagging suspicious ones for further investigation. Their system doesn't just look at the claim itself. It considers data from social media, weather reports, satellite imagery, and more. It has the power to uncover massive fraud rings that may have gone undetected for years.

Account takeover is where criminals gain access to your online accounts—e email, social media, or worse, your bank account. Security measures like passwords are increasingly ineffective against sophisticated hackers. That's where behavioral biometrics comes in. Companies like BioCatch use AI to analyze how you interact with your devices—everything from how you type to how you move your mouse. It creates a unique profile for each user, making it much harder for fraudsters to impersonate you.

AI is also helping businesses spot internal fraud. Employee fraud is a global problem, but it's notoriously hard to detect. That's where AI shines. It may analyze corporate spending data, flagging suspicious patterns that might indicate embezzlement or kickbacks.

Let's not forget about one of the most pernicious forms of fraud: money laundering. Traditionally, catching money launderers involved painstakingly tracing complex webs of transactions. Now, AI is doing the heavy lifting. HSBC partnered with Quantexa to develop an AI system that can analyze billions of transactions and external data points to spot potential money laundering.

Case Studies

JPMorgan Chase uses an AI system called COIN (Contract Intelligence) to speed up its legal document review process. This smart software can analyze and interpret thousands of complex legal documents in just seconds, a job that used to take lawyers around 360,000 hours each year. By using COIN,

JPMorgan Chase has drastically cut down the time and cost involved in legal reviews, while also improving the accuracy in spotting discrepancies and potential fraud in financial contracts.[2]

Danske Bank, facing high false positive rates in fraud detection, partnered with Teradata's Think Big Analytics team in 2016. They created an AI-driven fraud detection system that uses machine learning to analyze millions of transactions in real-time. This new system reduced false positives by 50 percent and increased fraud detection rates by 60 percent, providing actionable insights in under 300 milliseconds.[3]

---

1. Corporate Finance Institute. (n.d.). Quant fund. *Corporate Finance Institute.* Retrieved from https://corporatefinanceinstitute.com/resources/career-map/sell-side/capital-markets/quant-fund/

2. Topham, G. (2017, February 28). JP Morgan software does in seconds what took lawyers 360,000 hours. *The Independent.* Retrieved from https://www.independent.co.uk/news/business/news/jp-morgan-software-lawyers-coin-contract-intelligence-parsing-financial-deals-seconds-legal-working-hours-360000-a7603256.html

3. Teradata. (2017, October 23). Danske Bank and Teradata implement AI to monitor fraud. Retrieved from https://www.teradata.com/press-releases/2017/danske-bank-and-teradata-implement-ai

Part III

Strategic Implementation of AI

Chapter Ten

Building Your AI Strategy

This chapter is your roadmap to building a rock-solid AI strategy to catch up with and even beat the competition. You will learn the nitty-gritty of AI maturity—what it means, why it matters, and how to measure it. Then, you'll learn how to spot AI opportunities hiding in plain sight and align them with your business goals. We will also tackle the roadblocks that trip up even the savviest companies and show you how to leap over them.

Definition of AI Maturity and its Dimensions

AI maturity is a 'measure of how effectively an organization leverages AI technologies to drive business value. Let's break this down into some dimensions. First are strategy and vision. How deeply is AI integrated into a company's long-term plans? A mature organization doesn't just dabble in AI projects here and there but has a clear roadmap for how

AI will transform its business model. Next are data and infrastructure. You can't have mature AI without mature data practices. This dimension looks at how well a company collects, manages, and leverages its data. Then there's the talent and culture dimension. A company with high AI maturity fosters an AI-approving culture across the entire organization. This means upskilling existing employees, attracting top AI talent, and creating an environment where everyone understands and embraces AI.

Don't forget about the ethics and governance dimension. AI is becoming more powerful and mature organizations recognize the need for responsible AI practices like having clear policies on AI ethics, privacy, and bias mitigation.

Another significant dimension is AI operations or AIOps. Mature organizations have robust processes for testing, deploying, and monitoring AI systems. They're operationalizing AI models effectively. Companies like MLflow are at the forefront of this, providing tools to manage the entire machine-learning lifecycle.

Lastly, there is the value realization dimension. At the end of the day, AI maturity is about driving business value. Mature organizations have clear metrics for measuring the impact of their AI initiatives, and they're constantly iterating to improve ROI. They're not implementing AI for the sake of it but laser-focused on how it impacts the bottom line.

However, AI maturity isn't a one-size-fits-all concept. Different industries and companies will have different maturity profiles. A tech giant like Google will naturally have higher overall AI maturity than a local bakery. But that doesn't mean the bakery can't be mature in its context—maybe they're using AI to optimize their supply chain or predict daily sales with incredible accuracy.

The Stages of AI Maturity: From Initial Awareness to Transformational Use

AI maturing starts with the "Initial Awareness" stage. Companies first wake up to the potential of AI, fascinated by this shiny new toy but not quite sure what to do with it. At this stage, you might see a retail company experimenting with a simple chatbot on a website or a manufacturing firm trying basic predictive maintenance.

Next is the "Active Experimentation" stage. Here, companies start to take the technology more seriously, trying out different AI applications to see what sticks. You might see a bank testing out AI for fraud detection, or a healthcare provider piloting AI-powered diagnostic tools.

Once companies get more comfortable with AI, they enter the "Formalized Approach" stage, where things start to get more structured. Companies at this stage likely have a dedicated AI team and a more systematic approach to AI projects. They're actively looking for ways to integrate AI into their business processes. A great example is how UPS moved into this stage with the ORION (On-Road Integrated Optimization and Navigation) system, using AI to optimize delivery routes.

The next level is "Strategic Integration." At this point, AI is becoming a core part of the business strategy. Companies at this stage are using AI to drive significant business value and competitive advantage. These organizations reimage their entire business model around AI capabilities.

Finally, we reach the "Transformational Use" stage. This is AI maturity at its peak. Companies at this stage are being fundamentally transformed by it. It's like a master craftsman who's so good at their trade that they're redefining it. These companies are using AI to create entirely new business models, products, and markets. Think about how Tesla has used AI to reimagine what a car company can be—from autonomous driving to over-the-air updates that continuously improve vehicle performance.

Now, it's worth noting that these stages aren't always linear. Companies might be at different stages for different aspects of their business. For example, a bank might be at the "Transformational Use" stage for fraud detection but still at "Active Experimentation" for customer service AI. Also, the time it takes to move through these stages can vary wildly. Some tech-native companies might leap straight to "Transformational Use" in certain areas. Others, especially in more traditional industries, might spend years progressing through each stage.

Key Indicators of AI Maturity Within an Organization

Data infrastructure is the backbone of any mature AI operation. It cannot be just any data, but high-quality, accessible data that's ready for AI consumption. Companies with mature AI practices have robust data lakes or warehouses with clear data governance policies to easily access and analyze vast amounts of user data.

Another indicator is the presence of a cross-functional AI team. In mature organizations, AI isn't siloed in the IT department. Instead, you'll see data scientists, domain experts, and business leaders working together on AI initiatives. At the same time, continuous learning and upskilling programs, within IT teams and outside of them, are also a hallmark of AI-mature organizations. These companies recognize that AI is a rapidly evolving field and invest heavily in keeping their workforce up to date.

There is also the level of AI operationalization or MLOps. Mature organizations have easy processes for deploying and managing AI models at scale, getting them into production efficiently. The use of advanced AI techniques is also a good indicator of maturity. While less mature organizations might stick to basic predictive analytics, more advanced ones are likely to use techniques like deep learning, reinforcement learning, or exploring quantum machine learning. DeepMind, for example, uses advanced AI techniques like the AlphaFold system to predict protein structure.

Another sign is the level of AI integration across business processes. In mature organizations, AI is woven into the fabric of day-to-day operations. Likewise, mature organizations have clear metrics for evaluating the ROI of their AI initiatives. They can show exactly how it's improving efficiency, reducing costs, or driving revenue.

Aligning AI with Business Objectives

How can companies align artificial intelligence with their objectives? They should start with clear business objectives. Vague goals of implementing AI won't cut it. Instead, think specific. For instance, "reduce customer churn by 20 percent in the next year" or "increase supply chain efficiency by 15 percent." Once you have these concrete objectives, you can work backward to identify how AI can help achieve them.

Another aspect of alignment involves the right people. AI shouldn't be siloed in the IT department. To truly align with business objectives, you need cross-functional teams. Data scientists need to work together with domain experts and business leaders.

You should establish clear metrics for success. How will you know if your AI initiatives are moving the needle on your business objectives? UPS provides a great example here. When implementing the ORION system, they set targets like reducing fuel consumption, increasing delivery efficiency, and improving customer satisfaction. Moreover, continuous iteration needs to be a part of this goal. Business objectives can change, and your AI initiatives need to evolve with them. Spotify's initial objective was to provide personalized music recommendations, but as podcasts grew in popularity, they had to adapt AI algorithms quickly to include podcast recommendations as well.

It's also worth noting that aligning AI with business objectives means knowing when not to use AI. Not every problem needs an AI solution. For example, while AI could help inventory management, human greeters are still the best way to welcome customers. The bottom line is to use AI where it makes sense for your business goals, not just because it's trendy.

Identifying Opportunities for AI Integration

Many companies have integrated AI by using AI-powered chatbots, but the real opportunity lies in how you implement them. Look at Lemonade, the insurance company. They've created an AI assistant named Jim that handles claims in seconds. It processes claims end-to-end, verifies them against the policy, runs anti-fraud algorithms, and wires money to customers' bank accounts.

But AI isn't just for customer-facing roles. Let's discuss something less glamorous but just as necessary: inventory management. Walmart has been using AI to optimize its supply chain for years. The system can predict demand for over 500 million item-and-location combinations every week, avoiding stockouts and reducing waste, especially for perishables. Such AI integration can lead to significant cost savings and improved sustainability.

That's great for big companies, but what about smaller businesses? Consider a local bakery. It could use AI to analyze sales data and predict daily demand for different products to reduce waste and prevent popular items from selling out too early.

In healthcare, AI is opening up exciting possibilities. Researchers at Stanford University developed an AI system that identifies skin cancer with the same accuracy as board-certified dermatologists. It doesn't mean AI will replace doctors, but it could serve as a powerful screening tool that prioritizes cases and saves lives through earlier detection.

When identifying AI opportunities, look for repetitive, data-intensive tasks or tasks that require quick decision-making from multiple factors. For example, in agriculture, AI-powered sprayers can distinguish between crops and weeds and apply herbicides only where necessary. That way, they reduce chemical use, benefiting both farmers' bottom lines and the environment.

It's also worth considering how AI can enhance creativity and innovation. In the music industry, AI tools like AIVA (Artificial Intelligence Virtual Artist) compose original music for commercials, video games, and full-length albums. This software won't replace human musicians but opens new possibilities for collaboration between human and machine creativity.

Overcoming the Barriers to AI Adoption

One of the biggest hurdles to AI adoption is the skills gap. Many organizations simply don't have the in-house expertise to develop and deploy AI. One obvious solution is to partner with an expert and bring in external expertise while simultaneously training your staff. This approach solves the immediate shortage and builds long-term AI capabilities within the company.

Data quality and accessibility are another major stumbling block. AI models are only as good as the data they're trained on. This problem is particularly acute in sectors like healthcare, where data is often siloed and inconsistent, which incentivizes advocating for better data standardization and interoperability. Meanwhile, a whole industry has sprung up to help organizations clean, integrate, and manage data for AI.

Cost is also a barrier, especially for smaller organizations. But there are ways to deal with this. Automated machine learning allows companies to create custom machine learning models without needing a team of data scientists. Even traditional industries have found ways to take advantage of AI without breaking the bank, using it for tasks like inventory optimization.

Legal and ethical concerns also hold companies back. The fear of bias in AI is real and justified. To address this, open-source libraries exist to detect and mitigate bias in machine learning models. Similarly, resistance to change is a classic barrier in any technological adoption, and AI is no exception, especially given the ethical concerns. Some companies have overcome this by involving employees in the development of AI tools and demonstrating clear benefits, like faster problem-solving and reduced downtime.[1]

However, even once employees accept change, integration with existing systems can be a technical nightmare. Some industries, like manufacturing and energy, have solved this by creating "digital twins" before rolling new systems out, reducing risks and easing the integration process. Lastly, the "black box" nature of some AI systems can be a significant barrier, especially in regulated industries. To address this, there's been a push toward developing explainable AI models. In the financial sector, new approaches allow for understanding and explaining the factors behind AI-driven decisions to preserve regulatory compliance and customer trust.

Emerging Trends in AI and Machine Learning

One of the most talked-about trends is the rise of large language models (LLMs). These AI systems can engage in human-like conversations, translate languages, write articles, write code, and more. OpenAI's GPT-3 and its successor GPT-4 have been at the forefront of this trend. People have used them for everything, from drafting legal documents to creating marketing copy. But it's not just about bigger models. There's a growing focus on making AI more efficient and accessible. Google's BERT, a natural language processing model, has been compressed to run on smartphones. This trend toward "TinyML"—machine learning on small devices—opens up new possibilities for edge computing and IoT applications.

Another fascinating trend is the intersection of AI and scientific discovery. DeepMind's AlphaFold has made headlines by predicting protein structures with unprecedented accuracy, potentially revolutionizing drug discovery and our understanding of diseases. MIT researchers have also used machine learning to discover new antibiotic compounds, addressing the growing threat of antibiotic-resistant bacteria.

Explainable AI (XAI) is gaining traction since regulators demand more and more transparency in AI decision-making. IBM's AI Explainability 360 toolkit is a prime example. It gives developers a suite of algorithms to help explain AI model decisions. Similarly, the concept of "'AI for Good"' is significant. Microsoft's AI for Earth program partnered with organizations worldwide to apply AI to problems like wildlife conservation and climate change mitigation. One project uses computer vision to track elephant populations in Africa.

Federated learning is another trend worth watching. It allows AI models to be trained across multiple decentralized devices or servers without exchanging data samples, addressing privacy concerns in sensitive industries. Meanwhile, reinforcement learning, where AI agents learn by interacting with their environment, is making waves beyond game-playing. OpenAI's robot hand, which learned to manipulate a Rubik's Cube entirely through simulation, shows the potential for robotics and automation.

Multimodal AI, which can process and generate different types of data like text, images, and sound, is another exciting frontier. OpenAI's DALL-E and CLIP models, capable of generating images from text descriptions and understanding visual concepts from natural language, hint at a future where AI can seamlessly work across different modes of information.

Preparing for Disruption: How to Stay Ahead

It's not enough to send your team to a yearly conference and call it a day. You want to create an environment where people are genuinely excited to learn new things. Maybe it's setting up a monthly "show and tell" where they can share cool tech they've discovered or starting a company book club that tackles everything from AI ethics to sci-fi novels.

'If you're not failing sometimes, you're probably playing it too safe. The trick is to make those failures cheap and fast. Set up a system where teams can quickly test out new ideas without going through layers of bureaucracy. You're effectively giving everyone a lab to experiment in, with guardrails to prevent major explosions.

If it takes five meetings and a notarized letter from the CEO to try out a new tool, you may be in trouble. Consider accelerating this process through different means. Maybe it's rotating people through different departments or setting up cross-functional teams. The goal is to make your company as nimble as a startup, even if your company has been in the business for ages.

Data is useless if nobody knows how to refine it. Invest in making your team data-savvy. But keep in mind that this doesn't mean everyone needs to become a stats expert, but they should at least know how to ask the right questions and spot garbage data when they see it.

Keep an eye out for startups, universities, or even competitors you could team up with. Sometimes, the best ideas come from unexpected places, and you want to cast a wide net.

Remember, preparing for disruption also means solving problems in new ways. Keep your ears open to what your customers are griping about. Sometimes, the next big thing is hiding in a customer service complaint. Lastly, don't forget about your company culture. Create an environment where people feel safe to speak up, challenge the status quo, and occasionally break things (metaphorically, of course).

---

1. IBM Corporation. (n.d.). Improving risk management and compliance with AI. Retrieved from https://www.ibm.com/downloads/cas/GVAGA3JP

Chapter Eleven

Implementing AI in Your Business

This time, you'll learn how to set clear, actionable objectives that align with your business goals. We'll show you how to spot the perfect use cases for AI in your organization, even in places you might not expect.

Great ideas need great people to execute them. We'll tackle the challenge of building or acquiring the right talent and technology. And because AI is nothing without data, we'll explore the critical importance of data quality and governance. Finally, we will discuss the somewhat tricky but no less significant matter of AI compliance.

Defining Clear Objectives and Use Cases

First, objectives are your big-picture goals and the "why" behind your AI adoption. Maybe you're looking to boost efficiency, improve customer experience, or gain deeper insights from your data. Whatever it is, it needs to be tied directly to your business strategy and not feel overwhelming or unattainable. Consider a goal like "reduce customer service response time by 30 percent using AI-powered chatbots" or "increase supply chain efficiency by 20 percent through predictive analytics." These are measurable, time-bound objectives that give your team a clear target to aim for.

Once you've nailed down your objectives, it's time to identify use cases. These are the specific scenarios where AI can make a tangible difference in your business. When identifying use cases, start by looking at your pain points. Where are the bottlenecks in your processes? What tasks are eating up too much time or prone to human error? These are prime candidates for AI intervention. For example, if your finance team spends hours reconciling invoices, that's a potential use case for AI-powered document processing. But don't just focus on the obvious. Sometimes, the best use cases come from thinking outside the box. Could AI help you predict equipment failures before they happen? Or maybe it could analyze customer feedback to spot emerging trends. Brainstorm widely, then narrow down based on feasibility and potential impact.

Speaking of feasibility, it's easy to get caught up in sci-fi fantasies of what AI could do. But remember, we are still dealing with noticeable real-world constraints. Consider your available data, technical capabilities, and resources. A use case that requires data you don't have or skills your team doesn't possess is a non-starter. It's also important to prioritize your use cases. 'Look at the potential impact versus the effort required. Some use cases might offer quick wins with relatively little investment—these are great for building momentum and proving the value of AI to stakeholders.

Others might be more complex but offer transformative potential in the long run.

Build or Acquire the Right Talent and Technology

The AI field is hot right now, and finding the right people can feel like trying to catch lightning in a bottle. You're not just looking for data scientists and machine learning engineers. You need a mix of skills—people who can wrangle data, build models, and, most importantly, translate AI solutions into real business value. One approach is to grow your talent by investing in training programs for existing staff. Think of it as teaching everyone to speak a new language, not just hiring a few interpreters.

But let's face it, sometimes you need to bring in the experts. Hiring experienced AI professionals can jumpstart your efforts. The trick is to look beyond the buzzwords on resumes. Ideally, you want people who have successfully implemented AI solutions in real-world settings. However, given how young the field is, it's important to set realistic expectations. Furthermore, don't forget about the importance of domain expertise. An AI whiz who doesn't understand your industry might build a technically impressive solution that misses the mark business-wise.

The AI technology landscape is like a high-end candy store—many shiny options that all look tempting. Sometimes, this temptation is overwhelming, so you must take stock of what you already have. Many companies are sitting on a goldmine of data and computing resources they're not fully utilizing. Furthermore, services like Amazon SageMaker, Google Cloud AI, and Microsoft Azure AI offer powerful capabilities without the need for massive upfront investment. But beware of the "shiny object syndrome." Just because a technology is cutting-edge doesn't mean it's right for you. Sometimes, a well-implemented "boring" solution beats a half-baked attempt at the latest and greatest.

Finally, remember that open-source tools can be a game-changer, especially for companies just dipping their toes into AI. Frameworks like TensorFlow and PyTorch offer robust capabilities and have the added benefit of large, active communities.

The Importance of Data Quality and Data Governance

Think of data consumption like cooking. You can have a kitchen full of ingredients, but if half of them are spoiled, your meal's going to be poor. In AI terms, this means your data must be accurate, complete, consistent, and timely. The consequences of poor data quality can be pretty dramatic. Imagine an AI system making financial predictions based on data where someone accidentally shifted a decimal point. Suddenly, your million-dollar forecast becomes a ten-million-dollar blunder. Explaining that to the board will be neither easy nor pleasant.

On the positive side, good data quality can give you a competitive edge. When your AI models are trained on high-quality data, they will make better predictions, spot trends more accurately, and generally outperform your competitors' systems. So, how do you ensure data quality? It starts with good practices at the point of data collection. It could be implementing validation checks in your data entry systems, regularly auditing your databases, or using AI to spot anomalies in your data. It's an ongoing process, not a one-time cleanup.

If data quality is about fresh ingredients, data governance is about managing your entire kitchen. It sets the rules for data collection, storage, usage, and disposal. Good data governance isn't merely a matter of compliance, although that's certainly part of it. Rather, it cultivates a culture where data is a valuable asset, with clear policies about who can access what data, how to handle it, and what it can be used for. One aspect of data governance is data lineage—being able to trace where your data came from and how it changed. It helps you understand and explain your models' outputs. If your AI makes a decision, you need to be able to backtrack and see what data influenced that decision.

Another important element is data privacy and security. With regulations like GDPR and CCPA, mishandling personal data can lead to hefty fines. Beyond that, it's about maintaining trust with your customers and stakeholders. If word gets out that you're playing fast and loose with personal data, it can do serious damage to your brand.

Implementing good data governance isn't always easy. It requires changes to processes and sometimes even to company culture. You might need to invest in new tools for data cataloging, access control, or data quality monitoring. But think of it as an investment in your AI future.

Establish Governance, Ensuring Alignment with Business Strategy and Compliance with Regulation

Governance helps you steer your AI initiatives toward your business goals while keeping up with regulations and ethical considerations. It's easy to get caught up in the excitement of AI and start projects left and right. But if these projects don't align with your overall business objectives, you're essentially just experimenting with expensive trinkets. You need a framework to evaluate and prioritize AI initiatives based on their potential impact on your strategic goals. That might mean setting up an AI steering committee with representatives from different parts of the business who can determine if the AI projects address real business issues, not just pursuing technology for technology's sake.

Regarding compliance, the early AI may seem like the Wild West, but new regulations pop up constantly. There is the GDPR in Europe, CCPA in California, and a whole host of other regulations depending on your industry and location. Avoiding fines is not the only reason to watch out for compliance (although those can be hefty). Compliance also means maintaining trust with your customers and stakeholders. Nobody wants to be the company that makes headlines for misusing AI. So, your governance framework needs to include processes for regular audits of your AI systems to ensure they're playing by the rules.

Here's a familiar example: Google's Gemini AI faced significant backlash after pulling unverified, sometimes nonsensical, or parody content from sites like Reddit and displaying them in search results without proper context. Users reported bizarre answers, such as recommendations on "gluing cheese onto pizza" or discussions of eating rocks. For a company like Google, this kind of mistake can hurt its brand reputation, especially when users expect accurate information. It underscores the dire need for AI systems to distinguish between reliable and humorous content, bringing to light the significance of thorough testing and better implementation.

Chapter Twelve

Creating a Culture of Innovation

The secret of successful AI adoption is your people. You'll learn how to foster an AI-ready culture that embraces change instead of fearing it. We'll explore strategies to keep your team on the cutting edge with continuous learning and innovation. Furthermore, we will tackle the reskilling and upskilling of your workforce for the AI age. For the leaders, there's also the lowdown on what leadership looks like when AI enters the company.

Foster an AI-Ready Culture

To foster an AI-ready culture, you must tackle the fear factor first. Let's face it: when people hear "AI," many still think of Skynet or job-stealing robots. Your job is to flip that script and educate the team. Showcase how AI can augment human skills, not replace them. Consider AI hackathons, where teams compete to solve real business problems, or "AI for Dummies" lunch-and-learns, where complex concepts are broken down into bite-sized, relatable chunks. You'll want to create a safe space for experimentation, too. Remember, AI is all about learning from data, including mistakes. If your culture punishes every misstep, it will stifle innovation soon enough. Instead, set up sandboxes where teams can play with AI tools without fear of breaking something important.

Cross-functional collaboration is another piece of the puzzle. AI doesn't respect traditional departmental boundaries. You need data scientists talking to marketers, engineers chatting with sales representatives, and so forth. Maybe set up AI task forces that pull people from different departments to work on specific projects. Leadership buy-in is critical. If the C-suite isn't walking the talk, your AI initiatives are going to fall flat. Encourage leaders to use AI tools in their work and to share their experiences openly—both the wins and the "learning opportunities."

Now, let's talk about skills. An AI-ready culture is one where everyone has at least a basic understanding of AI concepts. This doesn't mean everyone needs to become a data scientist overnight. However, they should understand enough to see how AI could apply to their role. Think of it like digital literacy for the AI age. Consider setting up an AI mentorship program. Pair your AI experts with people from other departments. It's like a cultural exchange program, but instead of learning a new language, they're learning to speak the "language" of artificial intelligence.

Creating an Environment for Continuous Learning and Innovation

In AI, if you're not learning, you're falling behind. So, how do you keep your team up with new developments? Consider setting up an internal "AI Academy." It would include interactive, hands-on workshops, guest lectures from industry experts, and real-world project simulations. Maybe even throw in some gamification—a few things motivate learning, like a little friendly competition. Encourage your team to spend time exploring new AI tools and techniques. Google famously had its "20 percent time" policy, where employees could spend a fifth of their time on side projects. While they've moved away from that approach, the principle is solid. Give your people the space to play and experiment.

What's cutting-edge today might be old tomorrow, so you need to create an environment where new ideas can flourish. One approach is to establish an "AI lab" within your organization. It's a cross-functional group dedicated to exploring new AI applications that should have the freedom to fail fast and learn from those failures. Remember, in innovation, a failed experiment isn't a waste; it's valuable data.

Importantly, break down those silos. Set up regular "AI jam sessions" where people from different departments can brainstorm on how AI could solve business problems. Don't forget about external collaboration. Partner with universities, join AI consortiums or set up hackathons with other companies. Fresh perspectives can be the secret sauce that takes your AI initiatives from good to great.

Managing the Transition: Reskilling and Upskilling Strategies

Just like with any emerging technology, you don't need to understand the intricacies of how it works, but you do need to know how to use it effectively. Let's start with the basics. Everyone in your organization should have a fundamental understanding of AI concepts—what it is, what it can (and can't) do, and how it might impact each of their roles. They could learn in a series of workshops, online courses, or even an internal guide. But beyond the basics, your reskilling strategy should be role-specific. For instance, your marketing team might need to learn how to work with AI-powered analytics tools, while your customer service reps might need training on managing AI chatbots.

Here's a neat approach: create "AI apprenticeships" within your organization. Pair up AI experts with employees from different departments for hands-on learning. Don't forget about soft skills, either. After all, uniquely human skills like creativity, emotional intelligence, and complex problem-solving become even more valuable in an AI-driven world. Invest in developing these skills alongside technical ones to collaborate effectively with AI systems.

Your data scientists and engineers need to stay on the cutting edge of AI developments. Consider setting up an internal "AI dojo" where they can practice new techniques, experiment with the latest tools, and push the boundaries of what's possible. Reskilling and upskilling aren't a one-time thing, as the field of AI is evolving at breakneck speed. So, you need to create a culture of continuous learning. Here is a creative idea: set up an "AI challenge" program. Each month, present a real business problem and challenge teams to solve it using AI. It's like a hackathon but ongoing. This develops skills and fosters innovation and cross-departmental collaboration.

Leadership in the Age of AI

AI-era leaders need to embrace ambiguity like never before. The field is evolving almost too rapidly—frustrating if you're used to stability but exhilarating if you love a challenge. You don't need to be a technical expert, but you do need insatiable curiosity. It takes skill and dedication to ask the right questions.

Ethical leadership takes on a whole new dimension with AI. Leaders are now faced with ethical dilemmas never seen before. How do we prevent it from perpetuating biases? What about data privacy? This is where transparency comes in. AI can be a black box, but your leadership shouldn't be. Be open about AI initiatives, including the challenges and potential risks—you need to explain your decisions clearly to maintain trust.

Collaboration also takes on a new meaning in the AI age. Leaders need to be master bridge-builders, connecting departments that might never have worked together before. You might need data scientists brainstorming with HR or engineers collaborating with marketing. On that note, one fascinating challenge is managing human-AI collaboration. How do you create a culture where humans and AI systems work together seamlessly? You will have to play to everyone's strengths and create an environment where both can thrive. Leaders must be ready to pivot strategies, reallocate resources, and rethink entire business models at the drop of a hat.

Amidst all this change, some leadership fundamentals remain constant. You still need to inspire your team, set a clear vision, and make tough decisions. It's just that now, you're doing it in a world where AI is a key player, not just a tool.

Chapter Thirteen

The Ethical and Social Implications of AI in Business

This chapter demonstrates the complex interplay between AI's transformative potential and the ethical dilemmas it presents. We'll explore the dual-edged nature of AI technology, examining both its promising benefits and the challenges it poses to privacy, job security, and societal norms. The discussion will cover ethical principles for responsible AI use in business, the critical importance of data protection, and the changing dynamics of the workforce.

The Dual-Edged Nature of AI Technology: Potential Benefits and Ethical Dilemmas

On the bright side, AI has the potential to solve some of humanity's most pressing problems. Take healthcare. AI can analyze medical images and spot potential issues faster and more accurately than human doctors in many cases, catching cancers early, predicting heart disease, and helping develop new drugs. But here's where it gets tricky. The same AI that could save lives could also be used to invade privacy. Imagine an AI system that can predict your likelihood of developing a certain disease based on your social media posts or shopping habits. Sounds great for preventive care, but what if insurance companies got hold of that information? Suddenly, you might find yourself denied coverage because an AI decided you're a high risk. '""

In business, chatbots can handle customer queries 24/7, freeing up human staff for more complex issues, while predictive maintenance spots equipment failures before they happen, saving companies millions in downtime. However, this efficiency comes with its challenges. As AI takes over more routine tasks, what happens to the workers who used to do those jobs? Factory workers are already familiar with this concept, but AI is starting to encroach on white-collar jobs, too. It's great for productivity, but it could lead to significant job displacement if exploited.

Then, there's the issue of bias in AI systems. These systems learn from historical data, and if that data contains biases (which it often does), the AI will perpetuate and even amplify those biases. For example, an AI system used in hiring might discriminate against certain groups if it's trained on historical hiring data that reflects past discriminatory practices. It's like teaching a parrot to talk—it'll repeat whatever it hears, including the stuff you wish it wouldn't.

Privacy is another major concern. AI systems often require vast amounts of data to function effectively. But where does this data come from, and how is it being used? There's a fine line between personalized service and invasive surveillance. Speaking of data mining, the environmental impact of AI is also worth considering. Training large AI models requires significant computing power, which translates to high energy consumption, meaning carbon footprints the size of small countries in some cases. It's a bit ironic that some of the AI systems being developed to combat climate change might be contributing to the problem. On the flip side, AI could be a game-changer for environmental protection. From optimizing energy grids to tracking deforestation in real-time, AI has the potential to supercharge our conservation efforts.

Key Ethical Principles for AI Use in Business

Transparency is a cornerstone of ethical AI use in business. Organizations must be forthright about their AI systems, including what data they're processing and how they make decisions. An insurance firm might use AI to process claims, and it should clearly explain to customers how this system works, what factors it considers, and how it reaches its conclusions.

Fairness is another crucial principle. AI systems need to avoid bias and discrimination, particularly in areas like hiring, lending, and customer service. For example, if an AI is used in recruitment, it must be rigorously tested to prevent it from discriminating against candidates based on factors like gender, race, or age.

Privacy considerations are paramount when implementing AI in business. Companies must handle data ethically and in compliance with relevant regulations using techniques like differential privacy, which allows for data analysis while protecting individual privacy. A tech company could use this method to improve their products from user behavior without compromising personal information.

There should be clear lines of accountability and responsibility for AI decisions and their consequences. If an AI-powered chatbot were to malfunction and provide incorrect information to customers, the company should have protocols in place to quickly address the issue, take responsibility, and implement corrective measures.

While AI can process vast amounts of data and make recommendations, humans should always be involved in important decision-making processes. For example, in a financial services company, an AI might flag potentially fraudulent transactions, but a human analyst should review these flags before acting.

Systems need to be reliable and protected against manipulation or attacks via regular security audits, penetration testing, and continual updates to protect against new vulnerabilities. For a company using AI in its supply chain management, the system must be secure against cyber-attacks to maintain business continuity and protect sensitive data.

The broader societal impact of AI use in business must also be considered. This includes potential job displacement and environmental effects. A manufacturing company implementing AI and automation might invest in retraining programs for the workforce to help employees adapt to new roles that complement AI systems rather than being replaced.

Furthermore, businesses should be able to explain how their AI arrives at decisions, especially in fields like finance and healthcare. If an AI system is used to determine creditworthiness, the financial institution should be able to explain to customers what factors influenced the decision.

Beneficence, using AI for the benefit of individuals and society, is a fundamental ethical principle. In healthcare, an AI system might analyze medical images and assist in diagnosis. The primary goal should be improving patient outcomes and supporting healthcare professionals, not just increasing efficiency or profits.

Privacy and Data Protection

In 2020, the world created, captured, copied, and consumed an astounding 64.2 zettabytes of data.[1] To put that in perspective, if each gigabyte in a zettabyte were a brick, you could build 258 Great Walls of China, and this number is only growing. With all this data floating around, privacy concerns have naturally come to the forefront. People are increasingly aware of how their personal information is being used, and they're demanding better protection. This shift in public consciousness has led to the creation of laws like the European Union's General Data Protection Regulation (GDPR) and the California Consumer Privacy Act (CCPA). Under GDPR, companies can be fined up to €20 million or 4 percent of their global annual turnover, whichever is higher.[2] That's no small change, even for tech giants. And it's not just about avoiding fines—businesses are realizing that good privacy practices are good for their bottom line. Companies with strong privacy practices are more likely to retain customers and avoid data breaches.

Data breaches are a major concern for businesses using AI. In 2021, the average cost of a data breach reached $4.24 million, the highest in seventeen years.[3] AI systems, with their vast data requirements, can be particularly vulnerable.

But it's not all doom and gloom. AI itself is being used to enhance privacy and data protection. Differential privacy adds carefully calibrated noise to datasets, making it impossible to reverse-engineer individual data points while still allowing for accurate analysis, like adding a bit of static to a radio signal—you can still hear the music, but you can't pick out individual voices.

Another interesting development is the use of federated learning, which allows AI models to be trained on decentralized data. Instead of sending all the data to a central server, the model itself travels to where the data is stored, learns from it, and then moves on. Only the learnings, not the raw data, are shared.

Homomorphic encryption is another promising technology. It allows computations to be performed on encrypted data without decrypting it first. Imagine being able to ask someone to do your taxes without ever showing them your financial information – that's the power of homomorphic encryption.

Of course, these technical solutions aren't silver bullets. They need to be implemented thoughtfully and in conjunction with robust policies and procedures. Regular privacy impact assessments, clear data retention policies, and ongoing employee training are all components of a comprehensive privacy strategy. Even well-intentioned uses of data can have unintended consequences. AI systems can sometimes make unexpected connections, revealing sensitive information in surprising ways. That's why businesses must think carefully about what data they need to collect and how they're using it.

The Future of Work: AI and Automation

Repetitive tasks are prime candidates for automation. Think about assembly line work, data entry, or some aspects of customer service. Chatbots, for instance, are becoming increasingly sophisticated. They can handle basic inquiries, freeing up human agents to deal with more complex issues.

But it's not just blue-collar jobs feeling the heat. AI is making inroads into traditionally white-collar professions, like the legal field. AI-powered tools can review contracts faster and more accurately than junior lawyers. They're not replacing lawyers entirely, but they're certainly changing how legal work is done. In the medical field, AI is assisting with diagnostics and treatment plans. While it's not putting doctors out of work, it is augmenting their capabilities. Imagine an AI system that can analyze thousands of medical images in minutes, flagging potential issues for a radiologist to review.

Now, you might be thinking this all sounds a bit dystopian, but it's not all doom and gloom. Many of the ninety-seven million new roles will be in fields we can barely imagine today. Just as the internet created jobs like social media manager or SEO specialist, AI, and automation will spawn entirely new career paths. We're already seeing the emergence of roles like AI ethicists. There's also a growing demand for data scientists and machine learning engineers. And let's not forget the human skills that AI can't replicate yet—creativity, emotional intelligence, critical thinking. These "soft skills" are becoming increasingly valuable in the job market.

Of course, the transition won't be smooth for everyone. There's a very real risk of growing inequality if we don't manage it properly. Workers will need to upskill and reskill to stay relevant. This isn't just a task for individuals—businesses and governments have a role to play, too. Many companies are already investing heavily in employee training, focusing on those unique human capabilities that complement AI. Adaptability is becoming a prized trait in the workforce. Governments are also getting involved. Some countries are experimenting with ideas like universal basic income to cushion the impact of job displacement. Others are revamping their education systems better to prepare students for this new world of work.

Developing Ethical AI Strategies

So, what does an ethical AI strategy look like? It starts with clear principles. Many organizations are adopting frameworks based on concepts like fairness, accountability, transparency, and privacy. Take fairness, for instance. This isn't just about making sure your AI doesn't discriminate based on protected characteristics like race or gender; there are less obvious forms of bias. For example, a credit-scoring AI might inadvertently disadvantage people from lower-income areas, not because of their race, but because of where they live. Catching these subtle biases requires rigorous testing and diverse teams.

Transparency doesn't mean open-sourcing your entire AI model (though some companies do opt for that). Rather, you have to be clear about what your AI does, how it makes decisions, and what data it uses, particularly in high-stakes areas like healthcare or criminal justice. Imagine an AI system recommending sentences for criminal offenders. If we can't explain how it arrived at its recommendation, how can we see it as fair? Hence, accountability is closely tied to transparency. Someone needs to be responsible for the AI's decisions and actions by creating new roles like AI ethics officers or ethics boards. They need real power to influence how AI is developed and used.

Privacy is a huge concern in AI ethics. AI systems require vast amounts of data to function effectively, but data collection needs to be responsible. Differential privacy is one method, as it adds noise to datasets to protect individual privacy.

Questions arise about how to instill AI with human values and ethical principles. Should AI be programmed always to tell the truth, even if a "white lie" might be more socially appropriate? How do we handle cultural differences in ethical norms? These aren't just philosophical questions—they have real implications for how AI systems behave in the real world.

One interesting approach to ethics some companies are taking is the idea of "ethics by design"—baking ethical considerations into AI development from the very beginning rather than treating it as an afterthought. It's like how we now think about cybersecurity. It's not something you bolt on at the end but a fundamental part of the design process. Of course, developing an ethical AI strategy isn't a one-time thing. It needs to be continuously reviewed and updated as new ethical challenges emerge through audits of AI systems, ongoing staff training, and staying engaged with the wider conversation.

Emerging Trends and Ongoing Debates

One of the hottest trends right now is the rise of large language models (LLMs) like GPT-4. They generate human-like text, translate languages, write different kinds of creative content, and answer your questions in an informative way. They're being integrated into all sorts of business processes. Some companies use them to draft emails, generate marketing copy, or write code for their apps. However, this trend isn't without controversy. There are ongoing debates about the potential for these models to spread misinformation or malware, not to mention the thorny copyright issues that arise when AI generates content based on existing works.

Another emerging trend is AI in decision-making. We're seeing AI systems employed to make or assist with hiring, lending, and even criminal justice decisions. But this trend is sparking intense debate. Can these AI decision-makers be fair and unbiased and at what point does AI assistance in decision-making become AI replacement of human judgment?

The concept of "explainable AI" is gaining traction as a potential solution to some of these concerns. It involves developing AI systems that can explain their decisions in ways humans can understand and build trust. Imagine an AI system denying you a loan—wouldn't you want to know why? However, creating truly explainable AI is a significant technical challenge, especially for complex systems like neural networks.

Edge AI, running AI algorithms on local devices rather than in the cloud, is another trend that's picking up steam. It can offer reduced latency, improved privacy, and lower bandwidth usage. However, it also raises questions about the balance between edge and cloud computing and how to ensure consistency across distributed AI systems.

The debate around AI regulation is heating up, and there are growing calls for stronger oversight. The EU's proposed AI Act, which would classify AI systems based on their potential risk and impose stricter requirements on high-risk applications, is being closely watched. But there's no global consensus on how to regulate AI. Many people also rightfully argue that too much regulation could stifle innovation and the freedom of speech and expression, while others contend that it's necessary to prevent harm.

Another ongoing debate centers on the impact of AI on employment. While some studies predict significant job displacement due to AI and automation (a 2020 World Economic Forum report suggested 85 million jobs could be displaced by 2025), others argue that AI will create more jobs than it eliminates (the same report predicted ninety-seven million new roles). The reality is likely to be complex and varied across sectors and regions.[4]

The environmental impact of AI is also becoming a hot topic. Training large AI models can be incredibly energy-intensive. A single large AI model can emit as much carbon as five cars over their lifetimes. This has led to growing interest in "green AI"—developing more energy-efficient algorithms and using renewable energy for AI computation.

---

1. Statista. (2023). Volume of data/information created, captured, copied, and consumed worldwide from 2010 to 2025. Retrieved from https://www.statista.com/statistics/871513/worldwide-data-created
2. GDPR.eu. (n.d.). GDPR fines and penalties. Retrieved from https://gdpr-info.eu/issues/fines-penalties/
3. IBM Corporation. (2021, July 28). IBM report: Cost of a data breach hits record high during pandemic. Retrieved from https://newsroom.ibm.com/2021-07-28-IBM-Report-Cost-of-a-Data-Breach-Hits-Record-High-During-Pandemic

4. World Economic Forum. (2020, October 21). Recession and automation changes our future of work, but there are jobs coming, report says. Retrieved from https://www.weforum.org/press/2020/10/recession-and-automation-changes-our-future-of-work-but-there-are-jobs-coming-report-says-52c5162fce

Conclusion

It's time to wrap up our tour of AI in business, let's take a moment to reflect. We've covered a lot of ground, from the basics of AI technology to its far-reaching implications for the future of work and society. If your head is spinning a bit, don't worry, that's perfectly normal when grappling with a force as transformative as AI.

So, what are the key takeaways? First and foremost, AI is a fundamental shift in how we approach business problems. We've seen how AI can drive efficiency, enhance decision-making, and enable new products and services. But we've also seen that AI isn't a magic wand. It requires careful planning, robust data infrastructure, and a willingness to embrace change.

The success stories we've explored share common themes: strategic vision, strong data foundations, cross-functional collaboration, and a focus on solving real business problems. Retailers use AI to optimize their supply chain, reducing waste and improving profitability, while healthcare providers leverage AI to improve patient outcomes.

But we've also seen cautionary tales. Companies that rush to implement an AI system without considering bias in data risk leading to discriminatory outcomes. Organizations that invest heavily in AI without preparing their workforce may be faced with resistance and underutilization. These stories remind us that the path to AI adoption is not without pitfalls.

So, how do we overcome these challenges? On the technical front, start with good data. Invest in data infrastructure, enforce data quality standards, and build strong data governance practices. As for AI models, simpler isn't always worse—sometimes, a straightforward algorithm that your team understands is better than a complex black box.

Organizationally, focus on building AI literacy across your workforce. This isn't about turning everyone into data scientists but ensuring that employees at all levels understand AI's capabilities and limitations. Foster a culture of experimentation and continuous learning. And remember, AI adoption is as much about change management as it is about technology.

Societally, we must grapple with the ethical implications of AI. As we've discussed, issues of bias, privacy, and accountability aren't just PR concerns—they're fundamental to building AI systems that are fair, trustworthy, and beneficial to society. Having the most advanced technology isn't enough when building a sustainable competitive advantage.

One thing is clear: AI will play an increasingly important role in shaping companies. Preparing for this future means developing a long-term AI strategy that aligns with your business goals. It means investing in your people, fostering partnerships, and staying agile in the face of rapid technological change. Leaders must champion AI initiatives, allocate resources, and set the tone for ethical AI use. But more than that, they need to accommodate the human aspects—fears, visions, and a culture of innovation and continuous learning. So, stay curious, keep learning, and don't be afraid to experiment.

If you're hungry for more, there's a wealth of resources out there. Check out the latest research from academic institutions and tech companies. Follow AI thought leaders on social media. Join AI communities and attend conferences. And most importantly, experiment with AI in your organization.

The future of business is being written in lines of code, gigabytes of data, and artificial neurons. But it's not the AIs that are writing this future; it's us. It's leaders like you who will determine how we harness the power of AI to create value, solve problems, and shape a better world.

So, as you close this book and step back into your business, remember that the AI revolution is already here. The question is, are you ready to lead in this new era?

References

1. AIX. (2023). Case Study: Unilever's Integration of AI in the Supply Chain. AI Expert Network. Retrieved from https://aiexpert.network/case-study-unilevers-integration-of-ai-in-the-supply-chain

2. AIX. (2023). Case Study: Walmart's AI-Enhanced Supply Chain Operations. AI Expert Network. Retrieved from https://aiexpert.network/case-study-walmarts-ai-enhanced-supply-chain-operations

3. Boston Consulting Group. (2023). Predictive maintenance in manufacturing: Enhancing efficiency with AI. Retrieved from https://www.bcg.com/publications/2023/predicitive-maintenance-in-manufacturing

4. Chaturvedi, R., Verma, S., Das, R., & Dwivedi, Y. K. (2023). Social companionship with artificial intelligence: Recent trends

and future avenues. *Technological Forecasting and Social Change, 193,* 122634. https://doi.org/10.1016/j.techfore.2023.122634

5. Corporate Finance Institute. (n.d.). Quant fund. *Corporate Finance Institute.* Retrieved from https://corporatefinanceinstitute.com/resources/career-map/sell-side/capital-markets/quant-fund/

6. DigitalDefynd. (2024). 10 Successful AI Marketing Campaigns & Case Studies. DigitalDefynd. Retrieved from https://digitaldefynd.com/IQ/ai-marketing-campaigns/

7. GDPR.eu. (n.d.). GDPR fines and penalties. Retrieved from https://gdpr-info.eu/issues/fines-penalties/

8. Huang, M. H., & Rust, R. T. (2020). A strategic framework for artificial intelligence in marketing. *Journal of the Academy of Marketing Science, 49*(1), 30-50. https://doi.org/10.1007/s11747-020-00749-9

9. IBM Corporation. (n.d.). Improving risk management and compliance with AI. Retrieved from https://www.ibm.com/downloads/cas/GVAGA3JP

10. IBM Corporation. (n.d.). The business case for AI in HR. Retrieved from https://www.ibm.com/downloads/cas/A5YLEPBR

11. IBM Corporation. (2021, July 28). IBM report: Cost of a data breach hits record high during pandemic. Retrieved from https://newsroom.ibm.com/2021-07-28-IBM-Report-Cost-of-

a-Data-Breach-Hits-Record-High-During-Pandemic

12. McKinsey & Company. (2022). Tackling bias in artificial intelligence (and in humans). Retrieved from https://www.mckinsey.com/business-functions/mckinsey-analytics/our-insights/tackling-bias-in-artificial-intelligence-and-in-humans

13. McKinsey & Company. (2022). The state of AI in 2022—and a half decade in review. Retrieved from https://www.mckinsey.com/~/media/McKinsey/Business%20Functions/McKinsey%20Analytics/Our%20Insights/Why%20data%20culture%20matters/Why-data-culture-matters.ashx

14. McKinsey & Company. (2024). The state of AI in early 2024: Gen AI adoption spikes and starts to generate value. Retrieved from https://www.mckinsey.com/~/media/mckinsey/business%20functions/mckinsey%20analytics/our%20insights/the%20data%20driven%20enterprise%20of%202025/the-data-driven-enterprise-of-2025-final.pdf

15. National Institute of Standards and Technology. (2022). Towards a standard for identifying and managing bias in artificial intelligence (NIST Special Publication 1270). Retrieved from https://nvlpubs.nist.gov/nistpubs/SpecialPublications/NIST.SP.1270.pdf

16. Pew Research Center. (2023, April 20). AI in hiring and evaluating workers: What Americans think. *Pew Research Center:*

Internet & Technology. Retrieved from https://www.pewresearch.org/internet/2023/04/20/ai-in-hiring-and-evaluating-workers-what-americans-think/

17. Statista. (2023). Volume of data/information created, captured, copied, and consumed worldwide from 2010 to 2025. Retrieved from https://www.statista.com/statistics/871513/worldwide-data-created

18. Teradata. (2017, October 23). Danske Bank and Teradata implement AI to monitor fraud. Retrieved from https://www.teradata.com/press-releases/2017/danske-bank-and-teradata-implement-ai

19. Topham, G. (2017, February 28). JP Morgan software does in seconds what took lawyers 360,000 hours. *The Independent*. Retrieved from https://www.independent.co.uk/news/business/news/jp-morgan-software-lawyers-coin-contract-intelligence-parsing-financial-deals-seconds-legal-working-hours-360000-a7603256.html

20. World Economic Forum. (2020, October 21). Recession and automation changes our future of work, but there are jobs coming, report says. Retrieved from https://www.weforum.org/press/2020/10/recession-and-automation-changes-our-future-of-work-but-there-are-jobs-coming-report-says-52c5162fce